SCIENTISTS, EXPERTS, AND CIV

How do scientists, scholars, and other experts engage with the general public and with the communities affected by their work or residing in their sites of study? Where are the fine lines between public scholarship, civic engagement, and activism? Must academics "give back" once they collect data and publish results? In this volume, authors from a wide range of disciplines examine these relationships to assess how they can be fruitful or challenging. Describing the methodological and ethical issues that experts must consider when carrying out public scholarship, this book includes a checklist for critical factors of success in engagement and an examination of the role of digital social media in science communication. Illustrated by a range of case studies addressing environmental issues (climate change, resource use, post-disaster policy) and education, it offers an investigation into the levels and ways in which scholars can engage, and how and whether academics and experts who engage in community work and public scholarship are acknowledged and rewarded for doing so by their institutions. Also bringing into the debate the perspective of citizens who have collaborated with academics, the book offers an exploration of the democratizing potential of participatory action research.

Ashgate Studies in Environmental Policy and Practice

Series Editor: Adrian McDonald, University of Leeds, UK

Based on the Avebury Studies in Green Research series, this wide-ranging series still covers all aspects of research into environmental change and development. It will now focus primarily on environmental policy, management and implications (such as effects on agriculture, lifestyle, health etc.), and includes both innovative theoretical research and international practical case studies.

Also in the series

Communities in Transition: Protected Nature and
Local People in Eastern and Central Europe
Saska Petrova
ISBN 978 1 4094 4850 1

Sustainability and Short-term Policies
Improving Governance in Spatial Policy Interventions
Edited by Stefan Sjöblom, Kjell Andersson, Terry Marsden and Sarah Skerratt
ISBN 978 1 4094 4677 4

Energy Access, Poverty, and Development
The Governance of Small-Scale Renewable Energy in Developing Asia
Benjamin K. Sovacool and Ira Martina Drupady
ISBN 978 1 4094 4113 7

Tropical Wetland Management
The South-American Pantanal and the International Experience
Edited by Antonio Augusto Rossotto Ioris
ISBN 978 1 4094 1878 8

Rethinking Climate Change Research
Clean Technology, Culture and Communication
Edited by Pernille Almlund, Per Homann Jespersen and Søren Riis
ISBN 978 1 4094 2866 4

A New Agenda for Sustainability
Edited by Kurt Aagaard Nielsen, Bo Elling, Maria Figueroa and Erling Jelsøe
ISBN 978 0 7546 7976 9

Scientists, Experts, and Civic Engagement

Walking a Fine Line

AMY E. LESEN
Tulane University, New Orleans, USA

Routledge
Taylor & Francis Group

LONDON AND NEW YORK

First published 2015 by Ashgate Publishing

Published 2016 by Routledge
2 Park Square, Milton Park, Abingdon, Oxon OX14 4RN
711 Third Avenue, New York, NY 10017, USA

First issued in paperback 2018

Routledge is an imprint of the Taylor & Francis Group, an informa business

Copyright © Amy E. Lesen 2015

Amy E. Lesen has asserted her right under the Copyright, Designs and Patents Act, 1988, to be identified as the editor of this work.

British Library Cataloguing in Publication Data
A catalogue record for this book is available from the British Library

The Library of Congress Cataloging-in-Publication Data has been applied for
Scientists, experts, and civic engagement : walking a fine line / edited by Amy E. Lesen.
    pages cm. – (Ashgate studies in environmental policy and practice)
    Includes bibliographical references and index.
    ISBN 978–1–4724–1524–0 (hardback : alk. paper)
    1. Community and college—United States. 2. Learning and scholarship—Social aspects—United States. 3. Communication in learning and scholarship—Social aspects—United States. 4. Education, Higher—Social aspects—United States.
    5. Universities and colleges—United States—Public services.
    I. Lesen, Amy E.
    LC238.S45 2014
    378.1'03—dc23   2014021466

ISBN 13: 978-1-138-54660-8 (pbk)
ISBN 13: 978-1-4724-1524-0 (hbk)

For my mother, Ruth Weiser Lesen, 1940–1986,
my father, Edward J. Lesen,
and my stepmother, Clarice B. Pollock,
who taught me the importance of doing work that
improves the lives of my fellow citizens

Contents

List of Tables

Notes on Contributors

Richard Campanella, a geographer with the Tulane School of Architecture, is the author of numerous articles and seven critically acclaimed books on New Orleans, including *Bienville's Dilemma, Geographies of New Orleans*, and *Delta Urbanism*. The only two-time winner of the Louisiana Endowment for the Humanities' Book of the Year Award, Campanella has also received the Williams Prize for Louisiana History, the Mortar Board Award for Excellence in Teaching, and the Monroe Fellowship from Tulane's New Orleans Center for the Gulf South. He is also a monthly contributing writer on urban geography topics for the *The Times-Picayune, Preservation in Print* magazine, and the quarterly *Louisiana Cultural Vistas*.

Janice Cumberbatch PhD is a lecturer in the Centre for Resource Management and Environmental Studies, University of the West Indies, Cave Hill Campus, Barbados. She has a Doctorate in Participatory Planning, and teaches courses in Environmental Planning, Professional Skills for Natural Resource Management, Environmental Impact Assessment and Sustainable Tourism. Prior to joining CERMES, Dr. Cumberbatch was the founder and Executive Director of Social and Environmental Management Services Inc. (SEMS), a private consulting firm that offered services in the areas of social and environmental impact assessments and social surveys; social and environmental policy research; social and environmental planning; community mobilization; and seminar and workshop design and facilitation. Previously, Dr. Cumberbatch worked for the United Nations Children's Fund (UNICEF), was the Deputy Director of a local heritage NGO, the Barbados National Trust, and developed and coordinated the training portfolio for the Caribbean Natural Resources Institute (CANARI). Dr. Cumberbatch has over 20 years of professional experience in participatory research, and has worked on projects in many of the English-speaking islands including, Barbados, St. Lucia, Cayman Islands, Turks and Caicos Islands, Bermuda, Antigua, Anguilla, Grenada, St. Vincent, St. Kitts, Nevis, Trinidad, the British Virgin Islands and Jamaica.

Amy Koritz is Professor of English and Director of the Center for Civic Engagement at Drew University. She is the author of *Gendering Bodies/Performing Art* and *Culture Makers: Urban Performance and Literature in the 1920s*, and co-editor of *Civic Engagement in the Wake of Katrina*. Since 1998 she has worked to develop programs and courses for undergraduates that connect academic learning to the community beyond the university. At Drew, she has established the Civic Scholars Program, a scholarship program for students demonstrating excellence in community service; expanded community-based learning classes; and developed sustained, multi-disciplinary community-based projects that address environmental, educational, and economic issues facing underserved, minority, and immigrant communities. She serves on the National Advisory Board of Imagining America: Artists and Scholars in Public Life, and is Co-Principal Investigator of its Engaged Undergraduate Education Research Group.

Amy E. Lesen is Research Associate Professor in the Tulane-Xavier Center for Bioenvironmental Research at Tulane University in New Orleans. Lesen works on the coast and in urban estuaries. The overarching theme of her work is the interrelatedness between ecosystem function and human social dynamics in coastal cities and coastal communities, and how those systems are influenced by climate and environmental change. Most of her current work focuses on New Orleans and southeastern Louisiana. Lesen is also interested in the sociological aspects of biophysical science practice, doing research and writing about scientific public engagement, participatory research, and interdisciplinarity. Lesen has a Bachelor of Science degree from the University of Massachusetts at Amherst in Marine Fisheries Biology and a PhD from the University of California at Berkeley in Integrative Biology with a concentration in biological oceanography and paleoceanography. She was a faculty member at Dillard University in New Orleans from 2007 to 2014, years which encompassed the writing and development of this book.

Chief Albert P. Naquin is the Traditional Chief of the Isle de Jean Charles Band of Biloxi-Chitimacha-Choctaw Indians, located in Terrebonne Parish, Louisiana. A strong advocate for his people and homeland, he has represented his Tribe on numerous occasions at the State, Federal and National level including a visit to the United Nations in 2010. He has also traveled to Alaska to gain direction from their experience during the Exxon Valdez Spill, The Grandmothers Council in Montana, and was a representative for his Tribe at the state level for education. Chief Naquin is a retired Federal employee from the Department of Interior/Mineral Management Service (MMS). He was an oil field safety inspector in the Gulf of Mexico for MMS and for Bureau of Land Management (BLM) in Colorado and New Mexico. He is a Vietnam veteran and Ambassador for the Native Americans of the Louisiana Gulf Coast. He holds an Associate degree in Life Sciences from Nichols State University, is a gourd dancer, keeper and drummer on the Miracle Drum. He works with numerous local and national advocacy groups to bring about policy change that will bring progress not only for his Tribe but for indigenous people everywhere. He is proud to represent a people of such strength and follow the example of the many in his family who were Chief before him.

Margaret (Molly) Olsen is Professor and Chair in the Department of Hispanic and Latin American Studies at Macalester College in Saint Paul, Minnesota. Professor Olsen's research focuses on early modern transatlantic studies, as well as Caribbean literature and culture of the colonial and contemporary periods. More specifically, she explores the encounters between Spain, Africa and the Americas and how the conflicts of empire, colonialism, materialism and culture play out discursively in the sixteenth and seventeenth centuries. She is particularly interested in how Afro-Latino communities make strategic use of writing as well as other modes of expression—including orality and performance—to resist colonialism. Essential to her work is how colonial struggles continue to be relevant in our present historical context. Professor Olsen's publications include a book on Alonso de Sandoval's work called *Slavery and Salvation in Colonial Cartagena de Indias* (UP of Florida, 2004) and numerous articles on Afro-Latino discourse in the Americas and Spain that appear in journals that include *Hispanic Review*, *Revista de Estudios Hispánicos*, *Bulletin of the Comediantes*, *Revista Iberoamericana* and *Research in African Literatures*.

Kristina J. Peterson currently facilitates The Lowlander Center, a nonprofit organization that helps create solutions through education, research, and advocacy, beginning at the community level, for Lowland people and places in the bayous of Louisiana. Social and environmental justice is at the core of the Center's work. Through the Center, Lowlanders seek solutions to living with an ever-changing coastline and land loss, while visioning a future that builds capacity and resilience for place and people. Peterson's 30 years of post-disaster community redevelopment experience help communities to envision futures that mitigate vulnerabilities, to enhance existing systems, or to develop new systems to serve the public. Peterson was a founding board member of the National Hazards Mitigation Association, collaborator with FEMA's Project Impact, and national coordinator with Church World Service Disaster Services. She is an anthropologist who holds a PhD in Urban and Regional Planning from the University of New Orleans, a Master of Sacred Theology and a Master of Divinity from United Theological Seminary, and a Bachelor of Arts in Urban Studies-Ethnic Studies, University of Puget Sound. She was made a fellow in the Society of Applied Anthropology in 1998, received the Prince Award for outstanding dissertation, the William Gibson Environmental Award in 2010, and a citation from the State of Maryland for work on social and economic justice.

Stephen Tremaine is Bard College's Vice President for Early College Policies and Programs and is founding director of the Bard Early College in New Orleans, a tuition-free, satellite campus of Bard College for high school-aged students across New Orleans. Tremaine holds a BA from Bard College and an MA from Tulane University, both in literature. Tremaine is a New Orleans native and a graduate of the New Orleans Center for Creative Arts.

Preface

Amy E. Lesen

The Inspiration for This Book: Louisiana and Beyond

How do scientists, scholars, and other experts engage with the general public and the communities where we do our work and who are affected by our work? What is our responsibility to do so? How can we strive to do work that contributes to improving the lives of our fellow citizens?[1] Where are the fine lines between public scholarship, civic engagement, and activism? What are the considerations facing scholars and experts when our work puts us on various sides of those fine lines? When is the relationship between scholars and community members fruitful and successful, and when is it challenging? What are some of the methodological and ethical issues experts must consider when civically engaging and carrying out public scholarship? Are academics and experts who engage in community work and public scholarship acknowledged and rewarded for doing so by their institutions? What are the roles and responsibilities of academic institutions themselves in contributing to and participating in their surrounding communities? What is the seldom-heard perspective of citizens who have collaborated with academics?

These questions framed a National Science Foundation (NSF)-funded symposium I organized, with the collaboration of Richard Campanella and Julie Hernandez, in New Orleans in November 2010. We called the symposium *Walking a Fine Line: Scientists, Experts, and Civic Engagement* because we wanted to interrogate the conundrum facing scholars who step out of the academy and into the civic arena: they are crossing a boundary line that is often not explicitly acknowledged or well defined, but is policed by our colleagues, our institutions, the media, and societal attitudes. We were aware of the high costs that can come from crossing those borders—loss of respect by one's academic peers, becoming embroiled in public or media controversy, or even losing a job. We also wanted to explore the relationships that civically engaged scholars form with our fellow members of society—community members and community leaders, members of the press, elected officials, and policy- and decision-makers—because we were also cogent that in making and maintaining those relationships, we as academics must acknowledge and grapple with issues of power differentials, expertise, ownership of work, and mutual respect. The symposium brought together almost 30 scholars,

---

1   Here, and throughout all the chapters in this book, we do not use "citizen" as a political or legal term, but as a reference to a member of society.

thinkers, and academic rabble-rousers from New Orleans, greater Louisiana, across the United States and overseas. We were biophysical scientists, social scientists, activists, philosophers, and higher education professionals—and people whose work bridges those fields. The symposium was a great success, as we delved into these issues from many angles, discussing and debating among people we would never otherwise come into contact with in our day-to-day work or at our disciplinary conferences. The discourse was so rich that some months after the symposium, I asked the participants if they were interested in putting together a book, and, in addition to me, many of the authors in this volume are original participants from that symposium: Richard Campanella, Janice Cumberbatch, Amy Koritz, Margaret Molly Olsen, and Kristina Peterson.

I began thinking deeply about academic public engagement during the aftermath of Hurricane Katrina and the subsequent levee failures in New Orleans. When Katrina made landfall, I was a relatively new tenure-track Assistant Professor in New York City, two years out from finishing my PhD in coastal biology, and my work had become interdisciplinary in nature, encompassing issues of ecology, sustainability, and climate and environmental change in coastal cities and coastal communities. When the post-Katrina flooding happened in New Orleans (where I happened to also have many friends) I started thinking and reading about how vulnerable all coastal cities are to such disasters. As the weeks and months post-Katrina wore on, and I kept in close touch with my friends and colleagues from New Orleans who were spread all over the US waiting to be able to go back to their city and homes, it also became clear that New Orleans and the southeastern coast of Louisiana were now extremely popular places for scholars, researchers, and students from all over the world to study. But the scores of academics flocking to New Orleans—all doing important work—also made me think about the dilemmas this situation poses. What are the ethical implications when scholars come into a location—particularly one where people are in distress—study the situation, and then leave to go home and write an article or book for an academic audience? Isn't there a way this can be done where the research plan and the benefits of the work can be formulated with the intention of also benefitting the local residents? Or what about even collaborating with the local community? How is geography important here—whether the researcher herself is a local resident or an "outsider?"

Not surprisingly, I wasn't the only person thinking about these issues at that time, and the book *Civic Engagement in the Wake of Katrina* (2009), co-edited by Amy Koritz and with a chapter written by Richard Campanella (both contributors to this volume) is evidence of the rigorous attention that has been paid to the subject. But academic civic engagement is not just important and topical in Louisiana and New Orleans. This discussion is applicable and crucial in communities worldwide. The environmental and social dynamics currently at play on the US Gulf Coast are indeed fascinating and apt case studies for looking at academic civic engagement, and many of the authors in this volume work in southeastern Louisiana, or have in the past. But this is not a book about civic engagement in Louisiana, nor is it a "Katrina book." And although there is a heavy representation of authors who attended the *Fine Line* symposium, this book is also by no means a collection of conference proceedings: the chapters here reach beyond and in different directions than what any of us presented at that 2010 gathering. I am formally trained in biophysical science—though my current work crosses many disciplines—so I'm particularly interested in scientific civic engagement, and a number of the other authors in this volume include biophysical science in their discussions. Thus, "science" gained a starring role in the title of this book but as you'll see, the authors come from—and delve into—a wide range of academic disciplines spanning the biophysical and social sciences and the humanities.

Why Civic Engagement Now?

Humans in every locale are grappling with complex, global problems with numerous causes, widespread and often unpredictable effects, and impacts on both social and environmental systems. Addressing these problems will require the coordinated and combined efforts of—and effective communication between—biophysical scientists, social scientists and other scholars, policymakers, industry, and communities to solve, plan for, and remediate (Driscoll et al. 2012; Jasanoff 2010; Alexandrov et al. 2007). Involvement of private citizens and local communities in policy decisions is often deemed crucial (Watson 2005). Public discussion and inclusion of "local factors" is important in creating policy that is "socially acceptable" (Bray et al. 1997). Local cultural norms, local history, and local attitudes affect the values private citizens have and the decisions they make. The ability of a community to, for example, formulate policy on adaptation to climate change depends on "culture and place-specific characteristics that can be identified only through culture and place-specific research" (Adger 2003). These local, cultural, and place-specific considerations are often neglected when scientists, other scholars, and policy-makers (who very often do not live in the community where they are working, have no personal history there, and/or have little contact with many sectors of the local population) carry out their work. Policy tends to be more successful when developed by a collaborative group that includes "a relatively balanced mix of governmental and non-governmental participants" (Koontz and Johnson 2004). Many scholars of public policy advocate a democratic process that involves "ordinary" citizens (deLeon 1997; Fischer 1993). Academics are thus increasingly being called upon to do interdisciplinary and cross-disciplinary work to more effectively study and address these multifaceted phenomena that span environmental and social realms (Adger 2001; Backstrand 2003; Kahan et al. 2011; Lubchenco 1998; Maldonado 2012; Wiek et al. 2012). Scholars are also urged to engage and communicate more effectively with community members, policy makers, and other "stakeholders" (Casadevall and Fang 2012; Cortner 2000; Head 2007; Lesen 2012; Peters et al. 1999).

The nature of expertise and the social construction of knowledge (especially scientific knowledge) are topics that have occupied thinkers for many years

(Freudenburg, Frickel, and Gramling 1995; Latour 1987; Shrum 1984). Investigations into those questions have led to, and have been interwoven with, a wider discussion about the role that scientists, scholars and experts play in the social realm. Researchers and social commentators have studied the role of scientists, experts, and scholars in the civic arena, in communities, in policy-making and the workings of government; have problematized the idea of expert-as-objective-observer; and have often called on experts (especially scientists) to participate in scrutinizing these aspects of their own work and lives and to acknowledge the many associated ethical issues of how we carry out our work (Jasanoff 2002; 1995; 1994). Increasingly, scientists, scholars, and experts are no longer thought of as distant, objective observers and chroniclers of the environment, society and culture, but as humans whose work and methodologies are shaped by the world around them, whose work influences and is influenced by the civic arena, and who (due to the nature of their work) may sometimes be ethically compelled to play an active role in larger communities and the civic realm (Fischer 2000; Hildebrand 1955; Marx 1992; Stanton 2008).

Many of these concepts have been encompassed into the terms "expert, scholarly, or academic civic engagement." The topic is studied by scholars; universities are evaluating their faculty and students' roles in their communities; teachers from K-12 to the undergraduate and graduate levels are implementing service learning in their classrooms; and scholars of education are writing about community engagement and pedagogy. Universities have centers of civic engagement; it is now both a recognized methodology and a field of study (Gelmon et al. 2012; Jaeger, Jameson, and Clayton 2012; Schofer and Fourcade-Gourinchas 2001). Walter Kimbrough (who is the president of Dillard University, where I was a faculty member during the development and writing of this book) spoke with me in the autumn of 2013 about issues such as the efficacy of service learning and reward systems for faculty civic engagement:

> There should be a shared understanding of what [service learning] means ... What is the balance between serving the community and the students' learning? There needs to be a true value for the community with students also applying what they are learning. There is a difference between volunteerism and service learning ... Institutions have to look at how we reward people, lots of [higher education institutions] talk about valuing public service, but it's not built into the institutional value or reward system ... that is a conversation that needs to happen (Walter Kimbrough, personal communication).

Indeed, civic engagement in all its forms is now one of the most pressing concerns of many faculty, researchers, students, and institutions.

Our Framework: The Continuum of Engagement

During one of the first conversations I had with Richard Campanella, we agreed that there is a continuum of civic engagement. The authors in this volume delve deeply into or touch on every area on this continuum. The activities that are usually implicated in discussions of academic civic engagement encompass what has come to be called "service learning," in which a community-based project is incorporated into a university course curriculum; many universities now have a service learning requirement or a required number of "community service" hours to be completed, or, as I mentioned above, their own institutional centers for civic engagement. Amy Koritz, Margaret Molly Olsen, and Stephen Tremaine rigorously problematize many modes of civic engagement in higher education in their pieces. At another location on the spectrum of engagement are modes where the expert makes his or her work and results available to non-academics, for example through giving invited presentations about one's work to local community groups, or engaging with or as a participant in the popular media. Richard Campanella discusses his experiences with, and advises us about, those types of engagement in his chapter, pointing out that, in fact, public engagement through the media contains its own continuum of engagement from writing a column for a local or national popular publication, to opining about one's areas of expertise in an interview with a newspaper, or radio or television outlet. My own chapter addresses that area as well, delving into the specific topic of scientific civic engagement through digital social media. There is also a wide range of ways members of the academy—and policy-makers—can carry out participatory research or projects, and Janice Cumberbatch gives us a very comprehensive model and framework for all types of participation. Doing research and work that is inherently community-engaged and participatory from start-to-finish, and integrates collaboration throughout, is at the most immersive end of the civic engagement spectrum, and is where Kristina Peterson's inspiring work resides.

Save one, every author in this volume is a scholar who is or has been employed at a college or university, so the perspective of this book comes very much from the civically-engaged academy. But it was clear to me that this book would be incomplete and lacking in an important viewpoint if it did not include the voice of a community member—someone we posit as being "a member of the public"—who could tell us what scholarly public engagement has been like from that perspective. And so, one chapter here is the edited transcript of an interview and conversation with Chief Albert P. Naquin of the Isle de Jean Charles Band of Biloxi-Chitimacha-Choctaw Indians in coastal Terrebonne Parish in Louisiana. Chief Albert works closely with Kristina Peterson, and I have come to know Chief Albert well over the past two years. His community, facing numerous environmental challenges and "natural" and anthropogenic disasters now and over the past several years, has been a magnet for all types of outsiders who want to study them or who ostensibly want to help. Chief Albert and many in his surrounding area could write their own books about their vast experiences with well-meaning journalists, scholars, students, and officials. With his generous collaboration, this book includes Naquin's voice, experiences, wisdom, and expertise.

Part I, Civically Engaged Academicians: Theories, Challenges, and Opportunities includes the four chapters wherein each author goes into detail about his or her specific experiences with civic engagement, also including broader analysis of civic and public engagement as a whole: Campanella, Koritz, Olsen, and Tremaine. Each of these pieces also has an aspect of memoir. These four authors present some of the theory of civic engagement, their personal stories of engagement, and discussion of the role engagement has played in the trajectory of their careers. The four chapters in *Part II, How We Engage: Modes of Participation From Digital Social Media to Radical Democracy*, each give detailed treatment of civic engagement praxis, with Cumberbatch's all-inclusive framework for participation, Peterson's model for intensive Participatory Action Research (PAR), Naquin's experiences with engagement of all kinds, and my literature review about the use of Twitter and social media.

Conclusion

All the authors in this book are passionate about academic civic engagement and have made it the practice of their work, or their subject of study, or both. I think it is reasonable to say that we all want to contribute to creating an academy that is both responsive to and held responsible by its surrounding community, and that rewards scholars for being publicly engaged. We all interface or collaborate with non-academics in our work in ways that span the full continuum of possibilities for doing so. We all promote community participation in scholarship. Authored largely by people working within academia, this book is written primarily with our fellow scholars (including students at both the graduate and undergraduate level) in mind: our colleagues who desire to study or practice civic engagement, or who are already doing so and seek the viewpoint of their compatriots. But I also hope that this book will be both valuable and informative for those outside the academy who collaborate with academics, or who are looking for evidence of scholars wishing to do work that is useful and enriches lives of our fellow citizens. With this book, we hope to contribute to the growing body of scholarship and discussion about academic civic engagement and, in some small way, help speed along the establishment of civic engagement as an integral part of the way academia functions.

References

Adger, W.N. 2003. Social Capital, Collective Action, and Adaptation to Climate Change. *Economic Geography*, 79(4): 387–404.

Adger, W.N. 2001. Scales of governance and environmental justice for adaptation and mitigation of climate change. *Journal of International Development*, 13: 921–31.

Alexandrov, G.A., M. Heimann, C.D. Jones, and P. Tans. 2007. On 50th anniversary of the global carbon dioxide record. *Carbon Balance Management*, 2: 11–12.

Bäckstrand, K. 2003. Civic Science for Sustainability: Reframing the Role of Experts, Policy-Makers and Citizens in Environmental Governance. *Global Environmental Politics*, 3 (4): 24–41.

Bray, M., J. Hooke, and D. Carter. 1997. Planning for sea-level rise on the South Coast of England: advising the decision-makers. *Transactions of the Institute of British Geographers*, 22 (1): 13–30.

Casadevall, A., and F.C. Fang. 2012. Reforming Science: Methodological and Cultural Reforms. *Infection and Immunity*, 80(3): 891–6.

Cortner, H.J. 2000. Making science relevant to environment policy. *Environmental Science Policy*, 3: 21–30.

deLeon, P. 1997. *Democracy and the Policy Sciences*. Albany: SUNY Press.

Driscoll, C.T., K.F. Lambert, F.S. Chapin, D.J. Nowake, T.A. Spies, F.J. Swanson, D.B. Kittredge, and C.M. Hart. 2012. Science and Society: The Role of Long-term Studies in Environmental Stewardship. *Civil and Environmental Engineering*, Paper 7.

Fischer, Frank. 1993. Citizen Participation and the Democratization of Policy Expertise: From Theoretical Inquiry to Practical Cases. *Policy Sciences*, 26(3): 165–87.

Fischer, Frank. 2000. *Citizens, Experts, and the Environment: The Politics of Local Knowledge*. Durham, NC: Duke University Press.

Freudenburg, W.R., S. Frickel, and R. Gramling. 1995. Beyond the Nature/Society Divide: Learning to Think About a Mountain. *Sociological Forum*, 10 (3): 361–92.

Gelmon, S., L. Blanchard, K. Ryan, and S.D. Seifer. 2012. Building Capacity for Community-Engaged Scholarship: Evaluation of the Faculty Development Component of the Faculty for the Engaged Campus Initiative. *Journal of Higher Education Outreach and Engagement*, 16(1): 21–45.

Head, B.W. 2007. Community engagement: participation on whose terms? *Australian Journal of Political Science*, 42(3): 441–54.

Hildebrand, J.H. 1955. The Social Responsibility of Scientists. *Proceedings of the American Philosophical Society*, 99(2): 246–50.

Jaeger, A.J., J.K. Jameson, and P. Clayton. 2012. Institutionalization of Community-Engaged Scholarship at Institutions that are Both Land-Grant and Research Universities. *Journal of Higher Education Outreach and Engagement*, 16(1): 149–70.

Jasanoff, Sheila. 2010. Testing Time for Climate Science. *Science*, 328: 695–6.

Jasanoff, Sheila. 2002. Science and the Statistical Victim: Modernizing Knowledge in Breast Implant Litigation. *Social Studies of Science*, 32(1): 37–69.

Jasanoff, Sheila. 1995. Cooperation for What?: A View from the Sociological/ Cultural Study of Science Policy. *Social Studies of Science*, 25(2): 314–17.

Jasanoff, Sheila, ed. 1994. *Learning from Disaster: Risk Management after Bhopal*. Philadelphia: University of Pennsylvania Press.

Latour, B. 1987. *Science in action: How to follow scientists and engineers through society*. Harvard University Press.

Kahan, D.M., E. Peter, D. Braman, P. Slovic, M. Wittlin, O.L. Larrimore, and G. Mandel. 2011. The tragedy of the risk-perception commons: culture conflict, rationality conflict, and climate change. *Cultural Cognition Project Working Paper*, No. 89.

Koontz, Tomas M., and Elizabeth Moore Johnson. 2004. One Size Does Not Fit All: Matching Breadth of Stakeholder Participation to Watershed Group Accomplishments. *Policy Sciences*, 37(2): 185–204.

Koritz, Amy, and George J. Sanchez. 2009. *Civic engagement in the wake of Katrina*. Ann Arbor: University of Michigan Press.

Lesen, Amy E. 2012. Oil, floods, and fish: the social role of environmental scientists. *Journal of Environmental Studies and Sciences*, 2: 263–70.

Lubchenco, Jane. 1998. Entering the century of the environment: a new social contract for science. *Science*, 279: 491–7.

Maldonado, J. 2012. Climate Change and displacement: human rights and local knowledge as guiding principles for new policy initiatives. In *Climate change and fragile states: rethinking adaptation*, eds M. Hamza, and C. Corendea. Studies of the University: research, counsel, education, publication series of UNU-EHS, No. 16.

Marx, L. 1992. Environmental Degradation and the Ambiguous Social Role of Science and Technology. *Journal of the History of Biology*, 25(3): 449–68.

Peters, S.J., N.R. Jordan, and G. Lemme. 1999. Toward a public science: Building a new social contract between science and society. *Higher Education Exchange*, 6: 34–47.

Schofer, E., and M. Fourcade-Gourinchas. 2001. The Structural Contexts of Civic Engagement: Voluntary Association Membership in Comparative Perspective. *American Sociological Review*, 66(6): 806–28.

Shrum, S. 1984. Scientific Specialties and Technical Systems. *Social Studies of Science*, 14(1): 63–90.

Stanton, T.K. 2008. New times demand new scholarship: Opportunities and challenges for civic engagement at research universities. *Education, Citizenship and Social Justice*, 3(1): 19–42.

Watson, Robert T. 2005. Turning Science into Policy: Challenges and Experiences from the Science-Policy Interface. *Philosophical Transactions: Biological Sciences*, 360(1454): 471–7.

Wiek, A., F. Farioli, K. Fukushi, and M. Yarime. 2012 Sustainability science: bridging the gap between science and society. *Sustainability Science*, 7 (Supplement 1): 1–4.

Acknowledgments

I am grateful to the National Science Foundation (NSF) for Award #0924792 from the Science, Technology, and Society Program, which funded the November 2010 *Walking a Fine Line* symposium that inspired this book. Michael Gorman was my Program Officer during the grant period, and he was enthusiastic about the project and helpful throughout.

I would like to acknowledge all the participants of the symposium for rich intellectual fuel: Lovell Agwaramgbo, Doug Ahlers, Barbara Allen, Leonard Bahr, Michael Blum, Steve Buddington, Richard Campanella, Lee Clarke, Craig Colten, Janice Cumberbatch, Scott Frickel, Alexandra Giancarlo, Kevin Gotham, Morgan Grove, Jenny Hay, Julie Hernandez, Amy Koritz, Shirley Laska, Brian Mayer, Kelly Moore, Earthea A. Nance, Molly Olsen, Kristina Peterson, Allison Plyer, Lawrence Schell, John Sullivan, Bob Thomas, and Ivor van Heerden. I am especially grateful to Craig Colten who was involved in brainstorming the symposium and helping me write the grant proposal that funded it. Kelly Moore and John Sullivan were also key in the conception of this book.

Julie Hernandez was a wealth of ideas, energy, and fun in the coordination, planning, and execution of the conference.

Nadine Wallace was my right hand woman in helping me plan the symposium, and I value Nadine's friendship and wish her all the best as she continues on towards her bright future. Shawn Bosby was an angel and a Godsend in helping with logistics and travel for the symposium.

My colleagues at Dillard University, where I was based during this project, are a source of inspiration to me. My students and advisees are the reasons we do what we do every day, and their enthusiasm and bright futures give me hope. I am proud and honored to have been amongst the Dillard family from 2007 to summer 2014.

I am especially grateful to Abdalla Darwish for mentorship, support, and encouragement, and to Walter Kimbrough for taking the time to be interviewed during my research for this book.

I am honored to consider Chief Albert P. Naquin a friend, from whom I have learned so much, and I look forward to our continuing friendship and fruitful collaboration. I hope he has received from me even a small measure of the joy and knowledge I have gained from him.

Theresa Dardar of the Pointe-au-Chien Indian Tribe has also been a friend and valued collaborator.

Shirley Laska is a role model, my civic engagement guru, and a friend.

Kris Peterson continues to be my greatest teacher and inspiration in community engagement, and also a valued friend and treasured colleague. I thank her for hours of grant writing, discussing, confiding, eating good meals, hanging out with Chief Albert, and general rabble-rousing, and look forward to more of the same far into the future.

I am grateful to Katy Crossan at Ashgate, who was the first person I handed the book prospectus to, and who was enthusiastic, encouraging, and helpful throughout the process of making this book a reality.

I thank all the contributors to this book, who were each a joy to work with, and whose insights, musings, and analysis continue to be a source of learning for me. Stephen Tremaine deserves particular credit for agreeing to write—and delivering—a wonderful chapter relatively late in the process of the book's compilation.

I have the best friends and family anyone could ever hope for, including my life-changing community from the Michigan Womyn's Music Festival. You all know who you are.

This book would not exist without the collaboration of Richard Campanella, who encouraged me every step of the way from the moment I first called him on the phone to introduce myself, from advising me while I wrote the grant that eventually funded the symposium, to the planning and successful execution of the symposium, all the way through the conception and completion of this book.

I am grateful to Michael Martak for support and partnership during the years when all of this was blossoming.

My father, Edward Lesen, has been my cheerleader and "the best father in the whole world." Both he and my stepmother, Clarice Pollock, were my original role models for activism and civic engagement, and continue to be well into their 70s and 80s.

Cheryl Hedrick never fails to remind me who I am and what I am capable of doing, and supplies a good deal of the joy that keeps me going every day. Her patience, encouragement, understanding, and caretaking through the process of writing and editing this book was stellar.

Part I
Civically Engaged Academicians: Theories, Challenges, and Opportunities

Amy E. Lesen

The chapters that begin this book are thoughtful accounts of four scholars' voyages in civic engagement. All have a quality of memoir, and all trace the role civic engagement has played in their careers. Richard Campanella gives a firsthand account of, as he says in the opening to his chapter, the situation "when individual scholars, scientists, and experts emerge hesitantly from their archives and laboratories to *personally* answer calls from fellow citizens at a time of crisis." Campanella has done so since 2005 and this has propelled him from being what he has claimed was a somewhat obscure researcher to now being quite popular and well known amongst New Orleanians. In the process of recounting his journey, Campanella gives us keen insights into how and why we engage, and—something he often points out may be even more valuable—advises us to play close attention to those circumstances when one should *not* engage.

In her chapter, Amy Koritz, who was also introduced to civic engagement in New Orleans post-Katrina, deftly considers the status of civic engagement in higher education and how we arrived at this place, through the lens of her own academic career and her transformation from an Assistant Professor of English at Tulane University to currently Professor of English and Director of the Center for Civic Engagement at Drew University in New Jersey. Koritz investigates our institutional structures and frameworks of disciplinarity influence, as she says, how we "organize and deliver knowledge," and she discusses the effects these dynamics have on civic engagement in higher education. As a biophysical scientist who often collaborates with scholars outside my field, I appreciate the way Koritz contrasts the systems of knowledge and pedagogy between the sciences and the humanities, an analysis she sees as key to understanding higher education's current dominant model of civic engagement.

Margaret Olsen also delves deeply into the state of civic engagement in higher education, by giving us a detailed case study of her advanced Hispanic Studies course at Macalester College, in which community engagement and service learning are integral parts. She points out that incorporating civic engagement into teaching requires flexibility, and one of the most profound findings in her piece is, in her words, that "alternative pedagogies that promote democratic relationships are often antithetical to the structured classroom experience of modernity that higher education has replicated for two centuries and has only just begun to reconsider."

The relationship between civic engagement and democracy is a theme that recurs often in this book, and Olsen's positing of this in the framework of the classroom is an important contribution to this discourse and this book.

The combined ways civic engagement, education, and democracy are intertwined are also insightfully discussed in Stephen Tremaine's thorough analysis of The Bard Early College in New Orleans (BECNO), a program he founded that gives New Orleans high school students access to courses at a satellite campus of an elite American liberal arts college. I feel Tremaine's piece embodies a unique approach to thinking about civic engagement, as he forces us to turn inward and look at the educational process itself as engagement, coining a new term in the process: "civic enrollment." As he says:

> ... we are often accustomed to thinking [about] ... civic engagement in terms of the impact that engaged scholarship has on our students, on ourselves, and on the communities that we collaborate with. [We take] ... an unusual step toward unraveling these distinctions: rather than student, university, and community, Bard aims to collapse the three; rather than community service, Bard's ambition is toward community enrollment.

Education itself as civic engagement? A radical idea, and one I'm extremely pleased is included among the insights in this volume.

Chapter 1

"When You Leave Town, I'll Leave Town": Insights from a Civically Engaged Researcher in Postdiluvian New Orleans

Richard Campanella

It's no coincidence that the concept of expert civic engagement has about as many monikers as it has meanings. Its constituents may be swapped for any number of imprecise synonyms. The adjective "expert," for example, may be substituted with "researcher," "scholar," "scientist," or "academic;" "civic" could also be expressed as "social," "public," or "community;" and "engagement" might be called "service," "responsibility," "advocacy," or "participation." The resulting bundle of awkward and inexact phrases[1] all broadly imply a circumstance in which (1) individuals or institutions with recognized proficiency in key areas venture beyond their rarified research realms and (2) enter into the messy, contentious world of public affairs, toward (3) the resolution or mitigation of real-world problems. It's where theory meets practice, where knowledge production meets application, where the ivory tower opens its gate to the muddled masses—and where an academic can have more lasting influence in a single statement than in a thousand footnotes. It's also where words can be misconstrued, reputations sabotaged, careers derailed, and enemies made.

Readers will glean elsewhere in this volume perspectives on other forms of academic/civic interaction, such as service-learning programs aimed at involving students in local society, and community-based or participatory research, in which professors collaborate with lay stakeholders and share with them funding, findings, and credit. Readers will also hear from investigators of civic engagement as a social and epistemological phenomenon, or who evaluate its effectiveness in bringing about positive change. What I hope to explore in this chapter is the concept in its most literal form: when individual scholars, scientists, and experts emerge hesitantly from their archives and laboratories to *personally* answer calls from fellow citizens at a time of crisis. The intended audience is not those who are obligated or otherwise eager to step toward an open microphone, such as politicians, authorities, lobbyists, or activists. Rather I address those researchers, usually but

---

1 These include "public scholarship," "socially engaged science," "academic civic engagement," "civic expertise," "social entrepreneurship," and "applied research," among others.

not always affiliated with universities, who hesitate to do so—particularly those who employ scientific methods, who strive toward objectivity and dispassion, and who view an academy overtaken by advocacy as deleterious. Admittedly, my data source is perfectly *unscientific*—personal experiences over eight years—but given the nature of this topic, perhaps I may be forgiven for viewing my foray as something of a laboratory experiment in and of itself. It certainly was a learning experience, and I share my insights here.[2]

By way of background, I am a Tulane University-based geographer who for nearly 20 years has researched and written about the historical and present-day physical and human geography of greater New Orleans. Few people cared about this topic in years past, and those who did all generally knew each other. True, New Orleans has long attracted more public interest than similarly sized cities, but that curiosity had mostly been sated with light comprehensive overviews, "local color" literature, or romanticized fare detached from modern-day realities. With certain exceptions (jazz history, for example), critical scholarship on New Orleans was surprisingly spare, particularly in the perfectly unromantic discipline of geography.

All this changed when, on August 29, 2005, Hurricane Katrina struck. The levees breached, the metropolis flooded vastly and deeply, and people suffered and died in large numbers. I witnessed the catastrophe personally as well as through a professional lens, as the terrible events playing out before my eyes represented searing culminations of the very centuries-old geographical patterns and problems I and others had been documenting for years. Indeed, the Katrina deluge formed a real-time epilogue to a manuscript, five years in the making, I had submitted to my publisher just weeks earlier. Now, floodwaters rose, smoke plumes wafted above the cityscape, and desperation and lawlessness prevailed. After five days of stabilizing our own damaged (but unflooded) house and checking on neighbors and friends, we finally fled the apocalyptic conditions.

I've often been asked why an "expert" like me—a geographer no less, who supposedly knew all too well the vulnerability of bowl-shaped New Orleans to a hurricane like Katrina—chose to remain home rather than evacuate. Eventually I came to terms with the truth: I sought to bear witness to a city I loved at a momentous time in its history. In other words, the researcher in me trumped the responsible citizen (not to mention responsible spouse), and while I have since vowed never to make that decision again, bearing witness gave me valuable understanding into exactly what transpired during that terrible week. It also, frankly, gave me a certain level of "standing" among those who would later speak authoritatively about Katrina, not to mention priceless eye-witness content for future writing and lecturing. Thus my first insight: for all our lofty detachment,

---

2   I originally used the word "lessons" here and throughout this chapter, but came to find that word a bit too facile and didactic, perhaps even pedantic. Better to call them "insights," which accurately traces their origins to my personal situation, and suggests that their portability to the situations of others may be limited.

we researchers love and care about our subjects; objectivity and dispassion may be more in our heads than in our hearts. We're also professionals, and savvy ones at that. Some may even say opportunistic.

We eventually evacuated to Baton Rouge, where I began analyzing and writing about what had happened. Interrupting my research was a growing stream of inquiries from reporters, who tracked me down through the intricate grapevine of the New Orleans diaspora. Katrina had been making headline news day after day around the globe, and breathless journalists interviewed me about the city's complex history, society, and environment. New Orleans geography became hot; history became relevant; culture gained currency. Arcane topics few cared about previously suddenly became the talk of the nation—and the world, since New Orleans had long enjoyed international renown, and because the tragic news coming out of Louisiana arrived at a time when issues such as global warming, urban sustainability, social inequity, the role of government, and the competence of the Bush administration were all buzzing on the world media's radar screens.

New Orleanians, meanwhile, craved every bit of information they could find about their wrecked city, devouring the *Times-Picayune*'s heroic reporting and hanging on officials' every pronouncement. Neighbors and colleagues buttonholed me for advice and predictions, and invited me to the fringes of the unflooded region to give evening lectures, sometimes under flickering lights and military-enforced curfews. With maps and diagrams I explained the historical backstory of the failed levees, how people came to live in harm's way, and what needed to happen to prevent a recurrence—even as two additional Category-5 hurricanes (Rita and Wilma) struck or threatened Louisiana within two months of Katrina. These memorable evenings often concluded with heartfelt conversations about the very geophysical viability of this community, and more than a few people have told me with a wistful smile, "Well, when you leave town, I'll leave town." It was a heady and momentous time, and what I gleaned from it was this: Engage. Share your knowledge, your insights, your hunches, even if you find yourself in unchartered waters. You devoted immense amounts of time to understanding this topic, and while you may not know everything, you know more than most. Now your fellow citizens find themselves in a time of dire need, and call upon you to share your findings. Heed the call.

The nature of the media inquiries bespoke another realization. When a major news story breaks, journalists in their pursuit of expert sources find them most reliably via competitors' articles. They track down the cited names, interview them, and expose them to an even wider circuit of readers and source-seeking writers. Eventually, certain oft-cited experts find themselves on a permanent media Rolodex, and their media-based civic engagement grows not arithmetically but exponentially. While it eventually stabilizes as the crisis cools, rarely does it disappear, and oftentimes it resurfaces whenever the topic reenters the news cycle—like on the 29th of August in 2006 ("The First Anniversary"), in 2007, in 2008, 2009, 2010 ("The Fifth Anniversary"), 2011, 2012, and 2013, not to mention when hurricanes Rita, Gustav, Ike, Isaac, and Sandy struck. It's a lesson in the communicative power and sheer momentum of the modern world-media juggernaut. It's also a reminder that things can slip beyond the control of a cautious researcher once that power is tapped. Heed the call, I suggest, but don't make the call.

What escalated my engagement beyond lectures and interviews was a conversation I had with a weary flood victim in a coffee shop in October 2005. It started by happenstance, grew increasingly engrossing, and ended with his warm encouragement to participate in the post-disaster planning activities about to launch. So inspired, I ensconced myself for a few days and developed an empirical methodology toward guiding the controversial question of neighborhood recovery prioritization. The proposal made it onto the email circuit—this was before social media—and earned me an invitation to present to the Bring New Orleans Back Commission (BNOBC), the City Planning Commission, and other key forums. It next appeared as a guest editorial in the *Times-Picayune* on November 13, precisely as members of the influential Urban Land Institute (ULI) arrived on the behest of the BNOBC to advise on rebuilding. ULI officials read the editorial and called me into their closed-door sessions to apprise them of the city's history, geography, future, and exactly how the methodology would work.

Alas, the proposal was rigorously debated—but not adopted because, I learned afterwards, the ULI judged that certain key variables would have been too difficult to ascertain. Nevertheless, the experience proved to be a worthwhile one, as the proposal helped frame the public discourse on what was at stake while making the case for reason and rationality over politics and emotion. From a civic engagement standpoint, the proposal offered a *methodology* toward making a difficult decision, which is data-based, rather than the final decision itself, which is value-based. The *Times-Picayune* later described it as the first publicly proposed plan for determining the safest areas to rebuild (Carr 2005), and to this day I receive positive feedback on it. An insight I took away from this experience might seem rather prosaic, but it's proven valuable tactically: guest editorials rank among the finest ways for academics to opine publicly to the right audience, on *their* terms, in *their* words, without risk of misquotation or decontextualization. Since 2005 I've penned nearly a score to the *Times-Picayune* and other forums, and they remain one of my favorite ways to "walk the fine line" between research and engagement. Today I write a regular monthly column on civically relevant geographical topics for the *Times-Picayune*, as well as for a statewide architectural preservation magazine and a quarterly humanities journal, all of which I post on my personal website and point readers' attention to via social media.

Similarly, news articles on research results represent an optimal pathway for findings to disseminate publicly. In 2007, after the controversy over closing down low-lying heavily damaged neighborhoods ended with a resounding "no," I contemplated re-problematizing the argument from a negative to positive framing. Instead of talking about removing residents from lower areas, why not shift the conversation to maximizing the residential occupation of higher areas? Using the toolsets of geography—remote sensing, Geographic Information Systems, and map-making—my students identified and measured over 2000 open parcels sprinkled throughout the higher terrain near the Mississippi River. Using historical and recent demographic data, I determined that, if properly zoned, tens of thousands of residents could occupy those lands at a lower flood risk as well as a higher urban sustainability. I released the results in a white paper entitled "Above-Sea-Level New Orleans," which circulated widely on the Internet.[3] What caught the media's attention was not so much the numerical findings, but rather what I had thought was already well-understood—that half of New Orleans lay *above* sea level. Locals had been taught all their lives that their city was "below sea level," and most post-Katrina news reports unconditionally characterized New Orleans as such, which unnecessarily put the city at a disadvantage when arguing for rebuilding. My findings ended up on the front page of the *Times-Picayune* under the blaring headline, "HIGHER GROUND." Since then, educating the public about the strange history, dangers, and benefits of New Orleans' topographic elevation has become one of my favorite topics, and perhaps my premier contribution to the public understanding of this city.

An additional benefit of white papers and guest editorials is that they allow the contributor to control the nature and timing of the contribution. Herein I apply a paramount guideline for civic engagement: opine only when you have new data, insights, or perspectives to contribute to the discussion at hand. Don't rehash arguments already articulated; avoid merely exploiting your name recognition and institutional gravitas to advocate for your side of a controversy; and never rant or proselytize. Steer clear of sarcasm and cynicism, and assume the best of intentions among those with whom you disagree. That said, never lock horns with a lunatic: finger-jabbing militants, indignant pedants, or otherwise angry malcontents ought to be heard out, but not engaged. Think twice, too, about addressing a group of (understandably) distraught and highly emotional victims of the crisis at hand, for two reasons: they may not recognize the limits of your expertise (not to mention your authority), and may press you for the latest agency-level policy decisions which you are simply unable to deliver. Secondly, you may come across as aloof and "out of touch" (a favorite populist rebuke) or, worse, find yourself getting drawn into the passions of the moment. In the post-Katrina era I witnessed more than a few peers—geographers, planners, and environmental scientists—who advocated for the closure of heavily damaged low-lying subdivisions when surrounded by colleagues who concurred, but who nervously softened their message or demurred entirely when they found themselves face-to-face with incensed flood victims. I myself have felt this pressure, and it is intense. The civically engaged researcher is free to address either or both audiences with whatever recommendations, but the message must be consistent. If you're ambivalent, half-hearted, or just plain scared, stay home. Judicious engagement means knowing when *not* to engage.

---

3   Richard Campanella, *Above-Sea-Level New Orleans: The Residential Capacity of Orleans Parish's Higher Ground*, http://richcampanella.com/assets/pdf/study_Campanella%20analysis%20on%20Above-Sea-Level%20New%20Orleans.pdf, April 2007. Featured on front page of *Times-Picayune*, April 21, 2007.

One time a Dutch news crew interviewed me amid the wreckage of the Lower Ninth Ward on the question of geographical risk, on which, unfortunately, I had little to say that was reassuring. The crew members positioned me in a curiously specific spot, and I soon came to realize why: nearby, a homeowner whom they had interviewed earlier toiled stoically amid his ruins, and they had hoped to catch on camera a confrontation between a humorless researcher warning against rebuilding *vis-à-vis* a sympathetic flood victim struggling to do exactly that. I caught wind of the staging, ended the interview, and vowed not to subject myself to that sort of entrapment again—a vow that I've been forced to exercise a number of times since. Good television is rarely good civic engagement, and if you sense that you're being railroaded into a confrontation narrative, bail out. Relatedly, decline pressing last-minute media requests that seem rushed and frantic; chances are the journalists are using you to do their due diligence, else they seek nothing more than a glib synopsis or a quotable zinger just under deadline.

Beware also demagogues bearing microphones. In early 2006, a radio talk-show host scheduled me to speak about the results of my citywide research on the reopening of flooded businesses. Ongoing at the time was arguably the highest-stakes and closest-watched mayoral election in recent American history, to determine who would lead the postdiluvian city toward recovery. The radio show began with my reporting of empirical findings and observations of the trends and patterns at play. The host, however, had his eye on the election, and wanted me, "as an expert," to opine on which candidate would be better suited to abet business recovery. That wasn't part of my investigation, I explained. But, he pressed, politics is relevant to economics and surely it falls within the purview of a study of post-disaster business recovery. Having zero intention of parlaying my field investigation into a political statement, I politely declined the question again, this time with the sort of evasive pleasantries one hears on Sunday morning interview programs. "You sound just like a politician," he snapped, and the air went dead. How ironic, I later thought: my insistence to stick to the facts and remain politically neutral earned me the mark of a politician. I eventually developed a sixth sense for detecting and avoiding that sort of huffy talk-show populism.

In another case, two French journalists requested to interview me (once again in the Lower Ninth Ward, the favored backdrop of Katrina reporting to this day) about topics within my area of expertise. After a few obligatory factual queries on soil subsidence and coastal erosion, however, the interview shifted toward value-based questions and loaded political polemics. All were perfectly fair game for public discussion, but none needed my personal opining, and I deflected them with increasing unease. Luckily, the two chatterboxes spent most of the time answering their own questions, and I was able to keep my opinions to myself. Other media moments taught me to reject questions whose premises I disputed, to speak with brevity and coherence but never with oversimplification, to decline invitations to speculate or predict, to be comfortable with awkward silences and pregnant pauses (which interviewers use artfully to lure interviewees into provocative waters), and to avoid glibness, clichés, and trivialization. In short, media interviews are as fickle as they are potentially powerful, and should be handled with circumspection *because the journalist controls the moment.* Expert civic engagement is best deployed when the *expert* controls the moment, and that usually means writing or lecturing.

As editorials provide optimal conditions for written engagement, lectures do so for those of the spoken variety. A lecturer fairly well controls the content and tone of an evening of public education. To be sure, the Q&A session relinquishes some of that control, but the lecturer still retains the podium, and a skilled speaker can parley even oddball questions or rambling commentaries into high-quality public/researcher interaction. The recent popularity of TED conferences and Pecha Kucha events have brought form and refinement to the art of communicating complexity coherently through lectures, and both have successfully broken down barriers between experts and laypeople. Panel discussions, however, are a beast of a different color. A panelist never really knows where the discussion will go, particularly when the moderator is ineffective and the audience wants red meat. A meandering discussion means you may well be pressed to answer a question outside your area of expertise, and that's when the trouble begins. The audience expects everyone sitting at the dais to be proficient in this subject, and surrendering with a feeble "I'm sorry, that's outside my area ..." begs the question, "... so why are you up there?" My advice: Be prudent whenever lecturing on a controversial topic, but particularly when the venue is organized as a panel. Foresee sensitive and provocative questions, and be on the lookout for polarizing individuals poised to hurl invectives from the back of the room. (Unlike in 2005, you are now on YouTube). And always, always take the time to prepare. Event organizers often cheerfully dismiss any need for special preparation ("Oh, don't worry, just show up!"), but that is always bad advice. Organize your thoughts, get your facts straight, and "frontload" your insights and explanations so that, when the moment comes, you express them succinctly and articulately.

I have been using the words "academic," "researcher," "scientist," and "expert" more or less interchangeably in this chapter, but of course they are not synonymous. Some experts in the realms of politics, government, and private think-tanks (either partisan or nonpartisan) are professionally obligated to engage publicly, and gauge their success in part by their air time. Military, law enforcement, and corporate experts, on the other hand, may be institutionally circumscribed from sharing their views. Within the realm of science, those at the applied end are usually missioned to make their work relevant and their findings accessible, whereas those at the basic or "bench" end tend to look askance at any public and media contact, going so far as to ostracize peers who so engage.

Within the confines of the academy, too, engagement sensibilities vary. Experts at university-based research centers or institutes, which generally operate like think-tanks, are usually more amenable to civic participation than their department-based counterparts. Likewise, nontenured faculty such as adjuncts, research professors, or professors of practice generally engage with greater leeway than tenure-track assistant, associate, or full professors. (Interestingly, professors emeritus, liberated from peer disapproval by nature of their retirement, are known to become much more civically active in their fields than during their professorial days). Academics in the humanities and the social sciences, as well as those in professions such as law and architecture, are today much more comfortable with civic engagement compared to their predecessors, not to mention their counterparts in the physical sciences. Indeed, many progressive humanists reading this chapter might be perplexed and even appalled at the very suggestion that a "fine line" must be walked between research and advocacy; they see privileged academics as *obligated* to fight for an equitable and sustainable society. The counterargument holds that civic engagement distracts academics from their chief mission of knowledge production and dissemination, that it misallocates human and fiscal resources, that it threatens scientific objectivity, that it breezily presumes political-philosophical unanimity, and that other sectors (nonprofit, government, private, faith-based, or consortia such as the Union of Concerned Scientists or the National Academy of Science) already exist to address pressing social needs. The dichotomy, I believe, is not a false one; the concerns are legitimate. Both views may be reconciled not by forcing civic engagement upon academics, nor by banning it, but by capitalizing on the beneficial side and minimizing the detrimental. Hence this chapter, and this book.

The public, we hope, benefits from researcher civic engagement via edification and elucidation. Conversely, we would be disingenuous to presume that civically engaged researchers do not reap benefits of their own. My situation, though small and localized, serves as a case study: the public's level of interest in the geography of New Orleans skyrocketed after Hurricane Katrina, and my book sales and speaker invitations have risen commensurately. A colleague of mine who studies environmental communications jestingly referred to the phenomenon as "the Campanella effect"—that is, how media coverage of major news stories brings little-known researchers into the limelight (Thomas, 2010). Sociologist Shirly Laska, geographer Craig Colten, historian Lawrence Powell, criminologist Peter Sharf, demographer Allison Plyer, architect David Waggoner, environmental lawyers Oliver Houck and Mark Davis, and educator André Perry have all become oft-quoted voices in the post-Katrina civic landscape, to the great benefit of the public. All can also probably attest that media attention generates social capital for the researcher, which in turn converts to professional and sometimes fiscal capital, not to mention power and influence. For topics of national interest, the elevation can be professionally transformative, and not always in the right direction. Consider how coverage of the O.J. Simpson murder trial in 1994–1995 made over a little-known adjunct academic into an influential nightly cable-television brand named Greta Van Susteren. Or how the banking crisis of 2007–2008 raised the status of a Harvard professor named Elizabeth Warren to win a seat on the US Senate. Consider also the case of geologist Jack Shroder, whose identification of Osama Bin Laden's whereabouts based on the rocky terrain in the background of post-9/11 video got the University of Nebraska professor swept up into the worldwide whirlwind of what he later ruefully termed "media hyperbole" (Shroder 2005).

A few years later, Hurricane Katrina transformed a soft-spoken marine biologist from South Africa named Ivor Van Heerdan into an outspoken critic of the Army Corps of Engineers, as he became the first to demonstrate that levee failures and not the storm per se caused most of the flooding. The episode earned Van Heerdan hero status in New Orleans—but also the termination of a nontenured faculty appointment at a Louisiana State University-based research center.

Van Heerdan's case suggests that unease among the lords of academia, whether based on funding, elitism, or plain old jealousy, is one potential cost of civic engagement. Another is the time it takes away from traditional professional advancement: rarely does civic engagement of any sort count toward tenure, and it may well work against it. Costs are also incurred with the possibility of misspeaking, getting quoted out of context, or simply having a bad day, which increases with every media exposure and can easily derail a career. Then there are the criticisms, personal attacks, and allusions to violence (at least one of the above-mentioned individuals received a death threat) from unhappy members of the public. Expect it: if you take it upon yourself to engage publicly, the public has every right to engage right back and sneer at your presumed expertise—all to be preserved forever on the Internet and retrieved immediately via Google.

An engagement problematic also arises when journalists seek expert sources based on spatial rather than topical proximity. In 2010, for example, many Katrina-era New Orleans experts were pressed to weigh in on the BP oil spill, a disaster that was geographically nearby but strikingly different in every other way from the 2005 storm and flood. Knowing little about petroleum engineering and marine biology, I emphatically deferred to the latest crop of civically engaged experts, such as Tulane University's Ed Smith and Louisiana State University's Ed Overton, who experienced their own versions of "the Campanella effect." The University of New Orleans' Denise Reed and Loyola University's Robert Thomas are among those scientists who civically engaged successfully for both Katrina and the oil spill, and both continue to serve as compelling scholarly voices educating the public on coastal restoration.

When conditions stabilize, civically engaged researchers may find themselves curiously repositioned in the eyes of others, in terms of their community voice. Many laypeople in New Orleans, for example, have come to presume that my expertise in *some* topics necessarily qualifies me to opine expertly on *all* topics. I am constantly pressed by strangers inquiring about my "position" on such-and-such political issue, or whether I support an ordinance or policy or candidate. "I'm not running for office," I joke, to puzzled looks.

Whereas laypeople are prone to overestimate the expertise of experts and grant them excessive civic clout, others take the opposite track and question your basic constitutional right to opine simply as a citizen. You are no longer "John Doe, Littletown" or "Jane Doe, concerned citizen" if you wish to opine on a local issue; you remain "an expert on … at … University" because of the momentum of your prior expert engagement. Case in point: when a local think-tank whose staff I know and admire released a report on a rarely reported social problem in New Orleans,

I scoured their methodology and found their definitions to be suspiciously inclusive, and thus their findings likely overstated. I penned a letter to the editor (not a guest editorial) pointing this out to fellow readers, and signed it with name and address only, *sans* reference to affiliation or anything else. In response to *my* response, I received a string of tart emails from the lead author, chiding me for weighing in on a topic outside my area of expertise—despite that I had claimed none in my letter and opined merely as a citizen with constitutional rights.

In another case, I wrote an essay (not a research article) for an online journal on my perspectives as a resident of a neighborhood undergoing strident gentrification. With the help of social media—which was barely on the scene at the time of Katrina but has since transformed human interaction utterly—the piece "went viral," and generated by far the most polarized reaction of anything I have ever written; readers either enthusiastically embraced it or eviscerated it. The essay touched a nerve mostly because of the volatility of the topic of gentrification, but a portion of the passion, I suspect, can be traced to how the aura of "expertise" (unclaimed, incidentally) hovered over my words on this particular topic. I have since been invited to a number of venues to participate in panels, open-mike nights, and theatrical performances tackling gentrification. Given the volatility of this issue and the realization that I came bearing no further data or solutions, the invitations promised little more than a very unpleasant evening. "We'd love for you come!" I was warmly assured by the event organizers; "No special preparation! Just show up!"

No thanks, I replied. I have some research to do.

References

Campanella, R. 2007. *Above-Sea-Level New Orleans: The Residential Capacity of Orleans Parish's Higher Ground.* http://richcampanella.com/assets/pdf/study_Campanella%20analysis%20on%20Above-Sea-Level%20New%20Orleans.pdf.

Carr, M. 2005. Experts Include Science in Rebuilding Equation; Politics Noticeably Absent from Plan. *The Times-Picayune*, November 25, 1.

Shroder, J. 2005. Remote Sensing and GIS as Counterterrorism Tools in the Afghanistan war: Reality, Plus the Results of Media Hyperbole. *The Professional Geographer*, 57(4): 592–7.

Thomas, R. 2010. BP's Gulf Oil Gusher Attracts Bewildering Amounts of Media Coverage. But Where's the Depth? *SE Journal*, 20(3): 5–11.

Chapter 2

Beyond Pasteur's Quadrant: Science and the Liberal Arts in a Democracy

Amy Koritz

In May 2011, while leading a faculty seminar on teaching community-based learning courses, I learned two important facts about scientists who teach in undergraduate-serving, liberal-arts focused colleges. First, I learned that they feel obligated to sustain a curriculum that will give their students the best shot at doing well on the MCATs, regardless of how small the numbers of those students who are realistically competitive applicants for medical school. Secondly, I learned that these same professors are frustrated by the lack of cross-disciplinary collaboration that occurs among their colleagues. "If we can't do this at a liberal arts college," one commented, "where can it be done?" I have heard similar laments for most of my career. And it doesn't much matter whether the faculty member is at a liberal arts college or a research intensive university—the desire for cross-disciplinary collaboration seems to be trumped at every turn by the demands of a curriculum (and the administrative structures that instantiate it) anchored firmly in a department. Other ways of organizing knowledge and delivering courses become peripheral to this central core. Faculty time must be bought out in order to achieve anything other than a relatively linear and compartmentalized delivery of courses focused towards a major. Attempts to alter this status quo by faculty members themselves, by academic administrators, or by external organizations such as the American Association of Colleges and Universities run up against discipline-serving structures. These are, finally, where a faculty member finds an institutional home, where she will build a career as a teacher and researcher, and where—should she be one of the lucky few—her tenure will be decided.

This chapter explores these institutional questions, but institutions are, after all, made up of people, each of whom has made a series of choices that brought him or her to that place. Within the academy, these choices are constrained in ways that are often counterproductive to both the needs of our society and our students. Those who self-select for careers in academia are, in the first place, obviously good at school. They are, secondly, comfortable living in a world where ideas are at least as important as practical action, and where continuity is frequently valued more than innovation. Within this self-selected, self-replicating universe, expertise is defined as the ability to either use the content of one's discipline to discover new or different knowledge, or as giving one the ability to convey that content to the next generation of potential experts. In the liberal arts (as opposed to professional or technical fields), acting on any feelings of obligation or responsibility to communities beyond academia is pretty much optional. It should rather be central. So, while this chapter does speak to the role of science and its professors, the domain of its argument is broader. The entire higher education sector is changing rapidly, placing under threat the traditional trajectories we have relied on to build careers, and the professional identities they express. While the sciences have been thankfully immune from the loss of prestige and resources suffered by many other disciplines, it is hard to imagine a future where the sciences as well are not asked to reconsider their goals and priorities in undergraduate education.

Professional Identities and Narrative Trajectories

Each of us carries with us a story about who we are and how we got to be that person. For those of us in academia, that story includes a narrative about our professional identity and career trajectory—how we succeeded, why we failed, what our purposes are in doing what we do—and that narrative is deeply informed by our professional guild. As an English professor, my professional narrative included assumptions about what literature professors liked to do (read and write alone in offices), what they valued in fellow literary professionals (deep textual knowledge, conceptual complexity), and most importantly for the purposes of this essay, how we as a guild defined and recognized status and accomplishment. This narrative was inevitably entangled with my own sense of self. Thus, the first and only time I took the Myers-Briggs personality inventory I scored as an extreme introvert—at the hide-in-my-room-and-never-come-out end of the scale. Now, some 15 years later, when I tell people that I am introvert, they act surprised. I'm still an introvert, meaning that when I need to recharge my batteries I look for alone time rather than a party, but my relationship to the world has altered dramatically since that time.

In 2008 I left a faculty position to become the founding director of a Center for Civic Engagement at a small liberal arts college. In retrospect, this dramatic shift in my professional trajectory had a lot to do with an early and on-going discomfort with assumptions about the purposes and priorities of the literary professoriate. Perhaps because I did not major in the field as an undergraduate, or because I'm intellectually attracted to questions that cross disciplines, or because I'm just contrary that way, I found myself almost from the outset of my career pushing the boundaries of what my profession defined as its central activities and purposes. This uneasy relationship to literary studies helps make sense of the story I am about to tell; it is told from the perspective of someone who had to step outside the confines of her professional guild, and the identity it privileged, in order to do the work that seemed most urgently in need of doing. My purpose in this essay is to ask why that should be the case—and most particularly why higher education should support scientists willing to step outside their own guilds to shape new narratives of professional identity.

Here's my story. I was a traditional academic. I got a PhD in English and a tenure track job at Tulane University. I wrote a book, got tenure, and started to get involved in university governance and program development. I was into feminist and literary theory and really didn't pay much attention to where I was—New Orleans wasn't important in the academic world that determined my career. But I was genuinely interested in undergraduate students and what kind of education they were receiving.

Then a new provost came in who wanted to do something about our apparently unacceptable first-to-second year student retention rates. She organized a retreat. I was fourth or fifth on the wait-list for a faculty slot. Nobody else could (or perhaps, being more experienced than I, would) go. So there I was, a relatively newly minted associate professor at a resort, with lots of deans and a fancy consultant. Fast-forward four months and I am suddenly in charge of first-year experience programs. The first thing I did was research the topic—what did higher education researchers find made a difference in student persistence? It turns out that colleges retain their students at a higher rate when those students feel a strong connection to the place where they are attending college—not surprising. And here I was in this utterly compelling place—New Orleans. It seems nobody had thought to introduce our first-year students to anything beyond the French Quarter (which they found on their own). Luckily for me, at the same time I was learning about student retention, a colleague in Psychology was starting a service learning program. This was the late 1990s—the provost had to intervene to keep her nascent service-learning office from being tossed out of the department. One of her student-workers, though, recruited me to incorporate service-learning into a course for my first-year students, and I was introduced to the power and difficulty of community-engaged teaching and learning.

Now I had another problem. How can I both teach my discipline and teach service-learning classes? At this point in my professional trajectory, I spent a couple of years attempting to teach two different classes in the same space and time. My students and I would read literature and criticism and we would work with community organizations—mostly in school settings. Some semesters I would devote one day to traditional literary study and the other to community-based teamwork and reflection. My teaching felt schizophrenic. I also began to get frustrated that my classes seemed to be doing the same thing over and over again with our community partners. Were we really making any difference? I didn't know whether any other faculty members were working with the same partners, or, if so, whether their goals and approach were aligned with mine. It felt like I was providing a great experience for my students (my teaching evaluations were strong), but having little or no impact in a community that had both great need and great potential.

Hurricane Katrina was the last straw, but also a moment of clarity. I finally realized that I had to choose my priorities. Was my top priority passing on my discipline's view of its important content to the next generation of students? It was clearly important to me that what I had to offer as an educator had some relevance to the communities outside my university. So I focused on community impact; the literature part would have to take care of itself. And it did—perhaps not in the canonical sense of the discipline, but in the sense that there are authentic and relevant connections between work in the world and the narrative or other forms of representation that make up literary texts. Although these connections are not always easy to make, and to be frank, students often resist making them, their presence began to organize my approach to community-based learning. While in the humanities disciplines the work of forging and communicating these connections was challenging, the sciences and the social sciences seemed to me to have a path laid out for them—if only they would walk down it. I was envious.

The urgency of post-Katrina New Orleans made it easier to leave the traditional classroom behind. I wrote grants, developed partnerships with the Tulane/ Xavier Center for BioEnvironmental Research, with a cultural anthropologist at the Urban Institute, and with non-profits around the city. I embarked on a collaboration with arts faculty from Xavier, Dillard, and NYU to develop and teach a community arts course. Amazingly, this course is still being taught as a collaborative endeavor between Tulane Dance professor Barbara Haley and Xavier Art professor Ron Bechet. All of this opened up professional opportunities for me as well—publication, funding, conference participation. While my motives were primarily community-focused at this time, I began to notice that my undergraduate classes were attracting a cohort of students hungry for opportunities to connect classroom learning with real-world problem solving. Many of these students were English majors who had found that there was precious little in the department's standard offerings that spoke to their desire to have impact in the world. As I came to learn, however, researchers studying effective undergraduate education had good evidence that the kind of pedagogy I had stumbled into was indeed increasing student learning and persistence. What I was doing, that is, made sense in terms of the core mission of most colleges and universities in the United States—undergraduate education.

George Kuh, the well-known higher education researcher, has included service-learning and community-based learning among ten high-impact practices in undergraduate education. These practices have been shown to increase student retention and engagement. A meta-analysis of research conducted by Jayne E. Brownell and Lynn E. Swaner (2010) confirms the benefits of service learning and specifies program attributes that increase its impact. These include structured reflection, adequate oversight at the community site, meaningful work with the community partner, sufficient service hours to make student experience significant, and explicit connections made by the faculty member between the community work and the academic content of the class. I cannot over-emphasize this last point. One of my first observations when I first started teaching community-based learning classes was that my students resisted making these connections. In fact they seemed to have been trained to treat what happens in the classroom as an autonomous silo of knowledge and activity. We should push hard against this tendency, for several reasons:

- Real-world problems require the intellectual flexibility to accommodate perspectives from multiple disciplines and points of view. If you can't connect two explicitly related contexts—those of classroom and community in a community-based learning course—how will you be able to effectively engage in the kind of complex-problem-solving our globalized society needs?

- Employers are less interested in a student's major than in their skills. A survey released in April 2013 of employers conducted by Hart Research Associates for AAC&U indicates that capacity for innovation, critical thinking, clear communication, and complex problem-solving are more important than undergraduate major, and that ethical judgment and integrity, intercultural skills, and experience applying knowledge in real-world settings are high priorities.

- The science of learning supports the investment of higher education in community-based pedagogy. According to Dartmouth professor of Psychological and Brain Sciences Chris Jernstedt, the best learning environment for students is the most active one—the person who is doing is the person who is learning. His research indicates that effective learning is less about conveying content than about what students do with that content (Teagle Foundation Convening: Faculty Work and Student learning in the twenty-first Century, April 2013, New York City).

Clearly community-based learning is only one form of active learning, but it is a commonly practiced, extensively studied form that demonstrably helps higher education fulfill its core mission. From a purely practical, even instrumental, perspective, college classes that connect with real-world contexts, give students multiple opportunities to practice what they have learned in the classroom, and build "soft" skills such as communication and teamwork, benefit colleges' core constituency and source of income—students. In addition, as the National Taskforce on Civic Learning and Democratic Engagement (2012) reminded us in *A Crucible Moment: College Learning and Democracy's Future*, colleges also have a broader mission beyond preparing individuals for economically viable futures: educating the citizens of a democracy. Interestingly enough, however, members of the professoriate are like most people in that the behaviors that are rewarded are the ones they prioritize, rather than those supported by either idealistic mission statements or research-based evidence. Even if they wish to do so, institutions of higher education are limited in what they can do to change the behavior of professors, because they do not hold all the cards in the systems for distributing rewards. The professional guild of a professor's discipline is arguably more important than the institution in determining his career trajectory, and within the institution the department is the closest proxy available for this guild.

Silos and Citizens

Donald Stokes (1997) has offered a book-length argument for attenuating what had become a strict division between "pure" and "applied" research in the sciences.

Engaging in so-called basic science—research pursued for the sake of knowledge and discovery, divorced from any practical or commercial application—had become a measure of prestige in academic science. Stokes' goal was both to explain that this separation was historically and conceptually far less clear than the proponents of either extreme might wish and to explore the genuine benefits—social and scientific—of explicitly seeking to work in what he called "Pasteur's Quadrant." Rather than the relationship between pure and applied science (or to use Stokes' vocabulary, basic science and technology) being linear, with each situated at opposite ends of a straight line, he proposes that this relationship is best and most accurately represented using a quadrant model. In this model, space emerges for "use-inspired basic research." Without such a category, the work of a scientist such as Pasteur, who is committed both to understanding and to use, cannot be properly recognized and valued. Clearly the strict division between basic science and technology that Stokes decried is no longer nearly as powerful now. As money for basic science becomes scarcer, scientists must necessarily look to less disinterested funding sources to support their labs.

But my interest is less in the current state of funding for scientific research than in the very particular manifestation science education has in liberal arts focused undergraduate education. Stokes' insight is relevant here precisely because of its importance to how science is taught at the undergraduate level. How science is taught is in turn relevant to all of us in higher education because most undergraduate majors mimic the organization of knowledge modeled by the science disciplines. Students start out taking pretty much the same introductory courses and unless they successfully complete these—Biology, Chemistry, Organic Chemistry, etc.—they cannot move up to more specialized classes. In this respect, the educational model of the sciences is the master model—in economics, sociology, psychology, and perhaps to a lesser extent in the arts and humanities. Further, as science fields become increasingly complex and hyper-specialized, students are forced to commit earlier in their college careers to a major in the sciences or risk having to take on the expense of additional semesters in order to complete a program of study.

In this undergraduate science education is different only in degree from that offered at most colleges in most disciplines. As a whole, faculty members teaching undergraduates are considerably more focused on conveying the content knowledge of their disciplines than on either addressing the role of disciplinary knowledge in complex real-world problems or in conveying the purposes and values professionals in that discipline serve. Without an understanding of the public purposes of our disciplines, however, those disciplines slowly, inevitably, lose support and legitimacy in the eyes of the public. Perhaps more importantly for the sciences, without an on-going conversation about those purposes, the potential that they will shift in ways counter to the interests of a democratic society, and certainly counter to the interests of the less affluent members of that society, increases.

As Stokes points out, the sciences have always been able to secure public support, largely because of their ability to draw persuasive connections between basic research and technologic advances that serve non-scientific interests. The ability of scientists to link their research, however remotely, to national security, health, energy, and other larger goals embraced by decision-makers and their constituencies makes the social, economic, and political consequences of scientific research integral to the doing of science. The discomfort many scientists express when asked to make explicit these entanglements in the content of the curriculum offered to undergraduate majors is thus a conundrum to me, an outsider.

The distance between the social impact of science and the manner in which knowledge is structured and conveyed through the traditional major represents a distance between knowledge and practice in the academy that has made me uncomfortable since my graduate student days. This distance has been most effectively bridged in recent years in the turn towards workforce development as an end in itself for higher education. Threatening to drop completely out of the picture, then, is the connection between the teaching of disciplinary knowledge and our obligations to a good beyond the ambition and enrichment of the individual. Between that rock of knowledge for its own sake and the hard place of knowledge for remunerative practice alone is a third way of structuring and delivering an education that serves a greater good without ignoring either the pure or the practical motivations for its pursuit. Donald Stokes' name for this approach is Pasteur's quadrant. The philosopher and scholar of the professions, William Sullivan, would term it "civic professionalism."

Like Stokes, Sullivan takes issue with a bifurcated approach to knowledge that focuses on either what he calls the critical thinking or the instrumentalist agendas. As might be obvious, the critical thinking agenda elevates a distanced intellectual stance towards its content—analogous to the pure research quadrant in Stokes' model—while the instrumentalist agenda is analogous to the applied quadrant. Both Sullivan and Stokes are naming a felt absence in the organization of educational, and by extension professional, priorities. As Sullivan states the problem, "What seems particularly lacking in many institutions is serious effort to provide an integration of students' educational experiences with the orientation and resources necessary for the ethical application of knowledge as individuals, as workers, and as citizens and participants in civil society" (2012, 142–3). Granted, this is not quite the same as Pasteur's focus on integrating discovery with application, but it does, I would argue, speak to the same impulse. To serve a social need in the context of discovering new knowledge defines the scientist as a civic professional.

In *Democratic Professionalism*, Albert W. Dzur (2008) reminds us that sociology has traditionally defined a profession as having three components: specialized knowledge, the ability to regulate the practice of its own membership, and social responsibility: "The larger society can rightly expect that the influential knowledge and skills of professionals serve social purposes, especially since professions are granted significant leeway in regulating their own conduct" (Dzur 2008, 46). The implicit tension between the autonomy of professionals and their obligations to the greater good may explain in part why the latter (social purposes) aspect of this identity receives less attention in their education than the former (knowledge and skills). As a society we value autonomy highly, making the professions attractive. The imperative to serve social ends, however, undermines that attribute, and thus the prestige of those professionals who accede to this imperative. To be free from social demand—that is to work entirely in the realm of pure research, basic science—would seem to be an ideal solution to this tension. Except that then the case for self-regulation depends only on expertise—not social benefit. Without a claim to social benefit meanwhile, the rationale for public (i.e. taxpayer) support evaporates. But perhaps there is another way to think about social benefit. If a research scientist teaches medical students biochemistry (something my father did for most of his career), wouldn't that be providing sufficient social benefit to enable bench scientists to be left alone with their research? Perhaps, but only if funding to support that research placed no other expectations on them. So professionals do not get to make these decisions on their own—what counts as social benefit is in fact determined as much by representatives of the public being benefited as by the guild of the profession.

As professionals we can only ride the autonomy train so far. Given the pressures on higher education, in fact, the ride may be over for many of us. Seriously bending the cost curve for students probably means increasing workload, outsourcing knowledge transmission to the Khan Academy and Coursera, and kowtowing to the competency-based learning gods at Lumina and the Gates Foundations. These may all be good things—it is not my intention to judge. It is, however, my intention to raise the question: what now are the social benefits that underwrite the claim to autonomy of academic professionals? As I hinted above, the obvious answer, to anyone who reads *Inside Higher Education*, is workforce development. I really did not want to write that sentence. Nevertheless, to the extent that academics do not take seriously the task of persuading our constituencies that we also serve a higher purpose, this is where we are collectively heading—scientists and humanists alike. If you are reading this from a tenured perch in an elite private or flagship public college or university, of course, it will be easy to dismiss this claim. In fact, such institutions may not be headed in the same direction as others—so, for those particular readers, welcome to the one percent. For the rest of us, however, this might be a good time to consider our options.

There are many examples of science professors and professional organizations recognizing the importance of integrating disciplinary knowledge with both community need and cross-disciplinary problem-solving. In an eloquent essay on science and citizenship Matthew A. Fisher (2010), a Chemistry professor, notes that the American Chemical Society regularly calls for its members to include social benefit as a core component of the purposes of chemistry, but that his colleagues resist doing so. The newly revised MCAT exams reflect recognition on the part of the medical community that in fact a good doctor needs to know more than science. Acknowledging that medical outcomes are influenced by psychological and social factors, the MCAT now requires coursework in these fields. The symposium motivating this essay was replete with examples of scientists grappling with the demands and implications of their involvement, as scientists, with civil society. Once, however, scientists (or English professors) move outside their disciplinary comfort zone, their sense of their professional identity begins to shift. If this shift is dramatic enough, they may find their priorities altered and their insider status within their field threatened.

Medicine, unlike chemistry, is perhaps less a science than a science-using field of practice. But, so, one might argue, is most science education at the undergraduate level. This does not diminish the definition of an undergraduate science educator as a professional, any more than the same characterization of medicine would diminish the professional identity of a physician. Moreover, this discussion of professional identity misses the larger point. The social benefit provided by a physician is not the same as that provided by a research scientist. By the same token, neither is the social benefit provided by a science educator. Further, the purpose of an undergraduate education, in the sciences, or in any of the liberal arts disciplines, is only partially encompassed by the transmittal of disciplinary knowledge. As William Sullivan has argued, undergraduate liberal arts education should have as its goal neither the instrumentalist agenda specific to workforce development nor the critical thinking agenda specific to disciplinary knowledge, but the development of abilities that integrate knowledge, action, and values. This, after all, would be the terrain on which science-literate citizens would ideally make decisions impacting public policy, the environment, and a host of other areas. Sullivan objects to a view of liberal arts education that seeks only to produce disciplinary specialists. "Rather," he argues, "genuine liberal education entails a different understanding of the liberal arts teacher as an intensively discipline-using educator whose aim is not so much specialized knowledge as the fostering of practical wisdom" (2012, 155). Practical wisdom refers to the ability not only to learn the relevant content knowledge of a field, but to have the judgment to understand how to apply it in particular circumstances, and the moral compass to determine why is should or should not be used in a particular way in that time and place. The model of disciplinary education most dominant in higher education does not emphasize these goals.

Certainly most undergraduates are better thought of as future science-using-citizens than as nascent Nobel Prize winners. What, then is gained by a curricular model that, perhaps motivated by a desire to give all equal access to the career path of the researcher, ignores the development of integrative and practical skills that our citizens will need in order to make intelligent decisions about science-using (or abusing) public policies that undoubtedly were not covered in any of their college courses? At the very least, it would seem important to give students practice connecting the content knowledge of the sciences to other realms of knowledge and action, realms in which the norms of objectivity valued by scientists may not be taken for granted. If it is important that our graduates use knowledge gained in science classes in their daily lives at all, providing them the tools to do so will require more than a couple of introductory lab courses. This, then, is where the revised MCATs are a step in the right direction, but do not come close to going the distance. The approach is additive rather than integrative, and it ignores the importance of skills and knowledge in the formation of professionals that are not classroom based.

Higher education traditionally defines learning based on the credit hour. The credit hour, in turn, is a measure based on seat time in a classroom. Classroom learning, almost by definition, does not emphasize the relationship between knowledge and practice. Nor, unfortunately, does it pay much attention to the purposes and values motivating the kinds of knowledge covered, the practical uses of that knowledge in the world, and the social obligations of those credentialed experts deploying that knowledge. Given the recent surge in the acceptability of competency-based learning, in combination with the fact that STEM disciplines are those most likely to see traditional classroom teaching displaced by other forms of content delivery, science professors who primarily serve undergraduates should be particularly interested in other forms of pedagogy and understandings of expertise. Like most of my colleagues in the faculty, however, regardless of which liberal arts discipline they profess, many readers may reject suggestions that they alter their goals or approach in teaching their discipline on several grounds:

- There simply is not time to do more than cover the basic content of the discipline—how can these other goals possibly be shoehorned into an already crowded syllabus?
- As a college professor, why should I be expected to provide practical experience beyond my areas of expertise?
- The values and purposes of the discipline are implicit in everything taught—of course students will understand this aspect of the field.

At this point it seems important to relay a story told by one of participants in the symposium that gave rise to this volume. Apologizing in advance for any errors the distance of memory may have imposed on the narrative—I believe I've got the general thrust of the story right. A scientist was asked to serve on a committee of scientists with relevant expertise to review a controversial proposal by a governor to address an environmental challenge facing his state. Prior to the arrival of the governor's representatives, the scientists agreed amongst themselves that his proposal made no sense scientifically. The governor's representative appeared, only to announce that the decision had been made to move forward with the governor's proposal. Unlike English professors, whose expertise—however valuable—has little direct relevance to policy decisions by politicians, this group of experts was suddenly, and without warning, required to act in a context where power politics, not objective fact, was the currency. Given on-going controversies over climate change, evolution (still!), and other issues where politics can easily trump science in the public discourse, why is it that all science majors are not as a matter of course given some set of tools for responding to or engaging with social and political contexts that use science, perhaps irresponsibly, in making policy decisions?

As a faculty member who also directs a Center with the mission of connecting knowledge and action for the common good, my perspective on these issues is colored by a set of experiences and institutional expectations not shared by many of my faculty colleagues. I am, for example, much less beholden to the norms of my discipline, and more focused on community priorities. Nevertheless, I would venture the following responses to the rationales offered by the imaginary colleagues above:

- The coverage model is a black hole. It prioritizes factual knowledge at the expense of developing practical wisdom, and in this regard it can both never meet its own standards (there is always more knowledge, and it is always important), nor can it educate graduates equipped to deploy the insights of a discipline effectively in their decision-making as citizens.
- There is a reason why employers are looking for graduates who can work effectively in diverse teams. The work of the world occurs in concert with others. Find partners who know what you do not, and invite them into your space. Even better, bring yourself and your students into theirs.
- Unless your assessment of learning outcomes measures this implicitly conveyed material, how could you possibly know whether or not students understand it?

To elaborate on this last point, William Fisher notes that:

> ... most undergraduate science educators seem to assume that majors who have a basic understanding of the scientific concepts will automatically make connections between those concepts and global challenges such as HIV/AIDS, access to pharmaceutical drugs, or malnutrition ... Yet we see very little evidence that students make these connections when they are not an intentional part of the curriculum. (2010, 113)

In my ten plus years of teaching community-based learning classes, I have seldom seen students make connections between classroom and community contexts that they were not explicitly asked to make. Students learn to treat their classes as silos, and too frequently their professors do little to discourage this tendency.

Not only do students treat classes as silos, they also treat their entire college experience (at least in the liberal arts, traditional college-age contexts where I have taught) as something walled off from the constraints and responsibilities of the worlds of work, family, and community. It is little wonder, perhaps, that hiring managers and recent college graduates disagree profoundly in regards to the graduates' readiness to enter the workforce (Harris Interactive 2013). While preparation for work and career has received a large, and seemingly increasing, amount of attention since the Great Recession of 2008, those advocating for higher education as preparation for active engagement in the communities where graduates live and work struggle to be heard. This is not because those communities do not need science-literate citizens. When scientists are willing to engage with the research questions important to communities, they find enthusiastic partners. Consider, for example, the vibrant Science Shop movement in Europe. Science Shops are defined on the website of the international Science Shop network, Living Knowledge, as "small entities that carry out scientific research in a wide range of disciplines—usually free of charge and—on behalf of citizens and local civil society." These shops are beginning to emerge in the United States as well, in the form, for example, of the Midwest Knowledge Mobilization Network and the Public Science Project at CUNY Graduate Center directed by Michelle Fine and Maria Torre.

To what extent do the sciences as professional communities, and scientists as practitioners within those communities, believe in the public purposes of science and the importance of providing credible expertise that serves the needs of civil society—as defined not only by the scientists among us, but also by the non-scientists that make up the majority of our communities? Underlying much of the discussion regarding the public purposes of the disciplines in general, and the sciences in particular, is an implicit (or sometimes explicit) devaluation of any faculty work not characterized as research. Research, however, seems to the outsider to be defined as work considered important and interesting only within the confines of the professional guild. Its responsiveness to community-driven questions and concerns may be a nice side-effect of a research agenda, but few colleges and universities place this criterion anywhere close to the center in importance. That said, there has been a major effort underway, led by institutions such as Syracuse University and by organizations such as Community Campus Partnerships for Health and Imagining America, to re-envision tenure and promotion guidelines in ways that value public scholarship and the co-creation of knowledge with communities. This effort seems to be a constructive response to the on-going devaluation of liberal arts disciplines. If it is not clear to those outside the academy why what we do matters to them, what we value will cease to be supported. Higher education is in my view a public good, and those of us lucky enough to build careers in this sector should think carefully about the civic obligations our good fortune entails. It is a good time to become publicly engaged—if for no other reason than that you've got a tailwind built up by a generation of pioneering scholars.

New Narratives for Old Professions

At the start of this chapter I related the story of how I evolved from being an English professor at a research university to being the founding director of a Center for Civic Engagement at a liberal arts college. Embedded in that transformation were changes in my professional identity; these entailed changes in the story I told myself and others about my professional purposes and practices. I even speculated on the origins of or reasons for these changes. The impulse to create narratives that make sense of ourselves and our world is almost irresistible, and it turns out there are motives for this urge for narrative coherence that, as educators, we should attend to. In the first place, cognitive neuroscientists have determined that the brain is less like a warehouse—storing information to be retrieved—and more like an archeological dig. Our brain does not reflect or contain a map of reality; it gathers fragments of information and out of them constructs stories about reality (Teagle Foundation Convening: Faculty Work and Student learning in the twenty-first Century, April 2013, New York City). The role of narrative in how we understand our lives has been explored in depth by the developmental psychologist Jerome Bruner. He proposes that

> ... life narratives obviously reflect the prevailing theories about "possible lives" that are part of one's culture. Indeed, one important way of characterizing a culture is by the narrative models it makes available for describing the course of a life. And the tool kit of any culture is replete not only with a stock of canonical life narratives ... but with combinable formal constituents from which its members can construct their own life narratives. (2004, 694)

Besides being an attractive theory to anyone with an advanced degree in literary studies, Bruner's ideas resonate intuitively. Perhaps because my own life narrative combined some of the available formal constituents in relatively unusual ways, I was forced to be more self-conscious about the ways in which I explained my professional trajectory than many of my colleagues. Unfortunately, many of us in academe may soon see this need for new professional narratives forced on them for less benign reasons. As the institutional landscape undergoes tectonic shifts, as our students and their families become less able or willing to invest huge amounts of resources to pursue learning "for its own sake," as the debt with which higher education saddles its graduates becomes increasing unconscionable—despite the clear payoff for many in life-long earning—the traditional model driving our disciplines will become harder to sustain.

It may well be a good thing, then, that the canonical narratives available to academics in shaping and communicating a professional identity valued by our culture are sufficiently limited and limiting that expanding them may be a welcome by-product of those changes we cannot control or predict—but that are surely coming. Meanwhile, those of us committed to civic engagement as an integral part of our work and identity in academia struggle to locate and deploy those narratives that will give value and meaning to our work. For all of us, the genres of teaching, service, and research constrain our options. Not only does much community-engaged work integrate elements from all three categories, but the manner in which we work in all of them is too often parochial and narrowly focused. Faculty advisors in PhD programs continue to discourage graduate students from embarking on publicly engaged scholarship until after tenure (should they be so lucky). As higher education researcher Kerry Ann O'Meara (2011) has argued, the dominant paradigm for graduate education continues to produce new professors who expect their work to be predominately private, with an audience only of fellow professionals, and to be conducted in a context where individual autonomy rather than social benefit is most highly valued. This state of affairs does not engender a narrative with a happy ending, for society or for academic professionals.

Luckily for both groups, there are many other possible lives, possible pedagogies, possible research methodologies, and ways of asking questions that matter to science, scientists, students, and society. Let us hope that we do not need to leave the disciplines and institutions that have given us our professional homes in order to find the narratives that give our professional lives a public purpose.

References

Bruner, J. 2004. Life as narrative. *Social Research*, 7(3): 691–710.

Brownell, J.E. and L.E. Swaner. 2009. High-impact practices: applying the learning outcomes literature to the development of successful campus programs. *Peer Review*, 11(2): 26–30.

Dzur, A.W. 2008. *Democratic Professionalism.* University Park: Pennsylvania State University Press.

Fisher, M.A. 2010. Educating for scientific knowledge, awakening to a citizen's responsibility. In *Citizenship Across the Curriculum*, eds. M.B. Smith, R. S. Nowack, and L. Bernstein, 110–31. Bloomington: Indiana University Press.

Harris Interactive 2013. Bridge That Gap: Analyzing the Student Skill Index. http://www.insidehighered.com/sites/default/server_files/files/Bridge%20That%20Gap-v8.pdf.

Hart Research Associates. 2013. *It Takes More Than A Major: Employer Priorities for College Learning and Student Success.* http://www.aacu.org/leap/documents/2013_EmployerSurvey.pdf.

National Taskforce on Civic Learning and Democratic Engagement. 2012. *A Crucible Moment: College Learning and Democracy's Future.* Washington, D. C.: Association of American Colleges and Universities.

O'Meara, K.A. 2011. Faculty civic engagement: new training, assumptions, and markets needed for the engaged American scholar, in eds. J. Saltmarch and M. Hartley *"To Serve a Larger Purpose": Engagement for Democracy and the Transformation of Higher Education*, 177–98. Philadelphia: Temple University Press.

Stokes, D.E. 1997. *Pasteur's Quadrant: Basic Science and Technological Innovation.* Washington, D.C.: Brookings Institution Press.

Sullivan, W.M. 2012. Knowledge and judgment in practice as the twin aims of learning, in ed. D.W. Harward, *Transforming Undergraduate Education: Theory That Compels and Practices That Succeed*, 141–58. New York: Rowman and Littlefield.

Chapter 3

Somewhere Between the Ideal and the Real, the Civic Engagement "Expert" Learns and Lets Go

Margaret Molly Olsen

The incorporation of civic engagement and service learning into college courses presents a destabilization of the traditional classroom. While the word destabilization may connote for some an upheaval or undesirable shakeup of accepted protocol, in fact it is this very restructuring of the dynamics of power and knowledge that has made civic engagement such a fruitful—and indeed necessary—endeavor in contemporary higher education. And yet, challenges and pitfalls abound as we reimagine how student learning should unfold in relationship to society. This chapter explores the tension between idealized goals for civic engagement and the practical reality of engagement endeavors as illustrated by an advanced level culture course taught in Spanish at a small, urban liberal arts college in Saint Paul, Minnesota. The chapter begins by presenting a context of increased interest in student-centered learning in higher education and suggests that civic engagement entails even greater demands for a flexible pedagogy both in and out of the classroom. I then discuss why I value civic engagement in my own teaching and the significance I suggest it holds for all disciplines in our current historical moment, but particularly for the humanities. To illustrate some of the points I make, I dedicate a large part of the chapter to discussion of the advanced Hispanic Studies course I developed with the support of a fellowship from Project Pericles, a consortium of institutions that maintain explicit commitment to service learning.[1] Here, I delineate my learning objectives and civic engagement intentions for the course, all of which have inevitably transformed as I teach the

---

1 Project Pericles is a not-for-profit organization that encourages and facilitates commitments by colleges and universities to include social responsibility and participatory citizenship as essential elements of their educational programs. Founded in 2001 by philanthropist Eugene M. Lang, Project Pericles works directly with its member institutions as they individually and collaboratively develop model civic engagement programs in their classrooms, on their campuses, and in their communities. http://www.projectpericles.org/projectpericles/.

I would like to thank the Teagle Foundation, The Eugene M Lang Foundation, Project Pericles, Jan Liss and Pal Schadewald for their generous support of my civic engagement work at Macalester College.

class again, due as much to the shifting needs of nonprofits and other organizations as to student experiences. One of the points I make is that alternative pedagogies that promote democratic relationships are often antithetical to the structured classroom experience of modernity that higher education has replicated for two centuries and has only just begun to reconsider. I therefore hope to show that it is most sound to teach students to effectively navigate the challenges that arise in collaborative work with community organizations as one of the most essential learning objectives of any course that includes a civic engagement component.

In the traditional classroom setting, knowledge flows in a top-down fashion from the instructor to the student. Here, the focus lies on teaching and the best means by which the educator may communicate what s/he knows to a room full of students prepared to internalize the material. In such an environment, student attention is directed forward toward the teacher, often minimizing potentially productive interaction with classmates, and certainly diminishing the generative power of the student. One can easily visualize the traditional arrangement of desks aligned in perfect rows to face the front of the room, where the teacher is most often standing while the students are seated; where s/he is speaking while the students remain silent; and where s/he writes on the board as the students dutifully copy the salient points of the lecture. Such a model reflects Enlightenment values of order, structure and empirical creation of knowledge, values that founded our public school system and academy of higher education in the US. Despite its Enlightenment origins, however, in praxis it is not an especially democratic system, given that a single figure of authority, the teacher, holds the bulk of the power, manifested in the knowledge that s/he imparts to students who are more often than not passive receptacles of previous learning, rather than active producers of knowledge. How then, to achieve in the classroom a more democratic learning environment, while maintaining the integrity of the learning process?

Student-centered learning (SCL) and the learner-centered classroom are not new ideas, though the terms we now use to name the practices are. In the early twentieth century, educational philosopher John Dewey (1916) advocated a participatory, interactive learning process among students in US public education. Similarly in the global context, the Montessori and Waldorf educational philosophies, which focus on a holistic learning experience, also arose at the turn of the (twentieth) century and later made their way to the US. But over the past two decades, student-centered learning has gained a foothold in many US institutions of higher education. Using SCL, educators seek to cultivate an environment of collaborative learning in which students work with each other as well as with the professor in a joint creation and exchange of knowledge. The objective is not to cede control of the classroom or the trajectory of the course to students, but rather to provide a viable alternative to traditional, hierarchical imparting of knowledge, that is, an alternative model that seeks to empower the student as a learner (Jones 2007). Therefore, while the professor creates the syllabus, selects the reading list and determines the overarching objectives for the course, students are encouraged to contribute additional learning goals, participate actively in the learning/teaching process in dialogue with their classmates, and to develop paths of critical inquiry that respond to their academic interests and individual programs of study. In short, the decentering of the classroom allows the student the potential to become a producer of knowledge.

Of course, each classroom is distinct, and each professor must determine the degree to which his/her pedagogy will be—indeed, can be—student-centered. But by combining a rigorous framework of study with plenty of opportunity for student discovery and leadership, students and professor alike may enjoy the benefits of collaborative learning: vigorous critical debate, excitement of discovery, the creation of intellectual meaning and a sense of accomplishment, both personal and common. By creating the circumstances in which students can weave together various facets of their liberal arts education, or bring their study away experiences, employment or community work to bear on course learning objectives, for example, a professor helps to make learning more integrated and holistic for students. The seemingly disparate elements of students' education thus begin to make sense!

But professors must be wary: student-centered learning is rarely easy or less time-consuming than traditional pedagogy. By inviting students to undertake collaborative projects or interdisciplinary research, or encouraging students to responsibly lead class discussions, a professor more often than not is inviting more work for herself. Unless strict parameters are placed on assignments—and this gesture itself may contradict the freedom of discovery that SCL champions—individual or group projects can evolve in disparate ways that will require the professor to work in close tandem with each student. Conversely, if careful guidance in content is not provided, the quality of classroom instruction can quickly deteriorate. We have all witnessed in dismay as a student takes precious reading material and proceeds to butcher it and baffle fellow classmates in the process while leading a class discussion. "If only I had decided to present that particular text myself," the professor might think. We will return later to comment on this aspect of SCL.

Like student-centered learning, civic engagement and service learning have exploded in North American higher education courses over the past two decades. Clearly, both share a focus on practical learning and student initiative. Additionally, because more often than not students' civic engagement experiences will be unique and specific to the community sites where they are working, students ideally will need to take a primary, active role in creating meaning out of their experience as it relates to course material. Here again, the professor determines the theoretical and historical framework in which students can locate civic engagement labors, and will certainly serve as a guide to community practice, but ultimately, the student must be responsible for creating true meaning and understanding from the engagement. Later, we can look at some strategies that facilitate this process, and it will become clear how SCL and civic engagement are naturally complementary. We can already anticipate, though, that including a civic engagement component adds yet another layer of complexity to the pedagogical organization of a course. Our goal as educators is to make it a fruitful complexity.

But let us take a step back for a moment and consider what civic engagement, in its myriad expressions that are not able to be addressed in the chapter, contributes to student learning. Why should professors put forth the additional time and energy to make civic engagement a component of their teaching? What makes it valuable pedagogically? Most educators who embrace civic engagement are interested in two primary objectives: that students gain practical knowledge in their area of study, and that young scholars become responsible, participatory, ethical and compassionate citizens through meaningful collaborations in society. The Coalition for Civic Engagement and Leadership from the University of Maryland stipulates the following principles that motivate civic engagement (2005). Of course, not all of them need to be objectives for every course:

1. Learning from others, self, and environment to develop informed perspectives on social issues;
2. Recognizing and appreciating human diversity and commonality;
3. Behaving and working through controversy, with civility;
4. Participating actively in public life, public problem solving and community service;
5. Assuming leadership and membership roles in organizations;
6. Developing empathy, ethics, values and sense of social responsibility;
7. Promoting social justice locally and globally.

An additional point that was included in an earlier version of the CCEL definition merits inclusion in this chapter because it drives my own interest in civic engagement, which I will turn to briefly: Valuing diversity and building bridges across difference.

The ethical dimension of civic engagement is really the foundation of its practice, and by ethics I mean both the expectation that students will come to see themselves as deeply implicated in the political and socioeconomic processes of local and global societies, and in the critical formation of their relationship to communities different from their own. What is at stake, ultimately, is an enduring sense of the public good, which is essential to the survival of democracy. William M. Sullivan (2005), in his book *Work and Integrity* explains how for many centuries, artisans and skilled craftsmen gained a sense of the public good through apprenticeships. By working in close conjunction with a master or specialist in the field, apprentices came to appreciate their own ethical role in contributing to society at large. In the US, as many spheres of employment were professionalized in the mid- and late nineteenth century, specialized professional preparation was relegated to higher education. As a result, according to Sullivan, the deep ethical connection to society that apprenticeships had once offered began to fade at the same time that much of the practical, hands-on knowledge that they offered also diminished. For Sullivan, then, civic engagement is essential in higher education as a means to remedy the ethical and practical vacuum in which many students acquire knowledge in contemporary academe (196–201). In my own work, I share

Sullivan's belief that if education is divorced from the practical knowledge of real world problems, particularly our most pressing social justice concerns, our democracy is put at risk as students are increasingly unable to imagine their own professional labor as part of a greater social good that benefits public wellbeing.

Gregory Jay (2010), like Sullivan, sees the future of higher education, and the humanities in particular, as necessarily tied to a commitment to social justice issues and solidarity with the "public." But he warns that we must be ever-vigilant of how the public is conceptualized by higher education: while the academy traditionally viewed its contributions to the common good as a paternalistic gesture of "doing for," the social movements of the 1960s ruptured that hierarchy of knowledge and advanced an ideology of creation of knowledge through collaboration within the realm of a common, shared public. Jay insists that we must continue to develop a democratic understanding of these "new publics" in our civic engagement work, keeping in mind questions of power and access, and remembering that "not all authority need be academic" (21). By encouraging us to rethink who controls the production of knowledge, and to reconsider who belongs to "the public" and "the community," Jay advances a more universalized, inclusive and participatory definition of society that locates nonprofits and other community organizations as indispensable partners in an equitable relationship with academia. In other words, all engaged participants belong to the new public.

One sees, then, how ideals of student-centered learning and civic engagement relate to the public good and democracy, in no small part by valorizing knowledge that has previously been suppressed. The values that drive these recent pedagogical practices are centuries old and have shaped what we understand as a liberal education or liberal arts education. William Cronon's classic piece on liberal arts education "Only Connect" (1998) speaks to what we really mean in the United States when we speak of the *artes liberales*, and he identifies freedom as integral to our understanding of liberal education:

> There was nothing vague about the *artes liberales*. They were a very concrete list of seven subjects: the *trivium*, which consisted of grammar, logic, and rhetoric; and the *quadrivium*, which consisted of arithmetic, geometry, astronomy, and music. Together, these were the forms of knowledge worthy of a free man. We should remember the powerful class and gender biases that were built into this vision of freedom. The "free men" who studied the liberal arts were male aristocrats; these specialized bodies of knowledge were status markers that set them apart from "unfree" serfs and peasants, as well as from the members of other vulgar and ignoble classes. (2)

Of course, we are more sensitive now to equality of access to education in the US, even if we still have a long way to go to achieve it in praxis. In fact, my own interest in civic engagement is fueled by the desperate need for student awareness of how our trajectory toward equal access to resources and opportunity in this country has been derailed over the past 30 years of neoliberal economics.

I agree with Cronon that a liberal arts education should "nurture human freedom and growth," but I believe that it should not merely nourish the freedom and growth of the individual who is fortunate enough to attend college (Jay 2010, 11). Rather, I insist that we as citizens should be consistently working to promote these values for all members of society, and clearly as part of that labor, we should reinvigorate our commitment to universal education at all levels. In my courses, civic engagement is ideally a means by which collaboration between students and local communities can help fulfill that goal.

The course I would like to share as a means to illustrate how one might navigate the tension between ideal objectives and the real experiences of civic engagement is an upper-level course I designed for the Hispanic and Latin American Studies Department at Macalester College called Cultural Survival: Resisting the Legacy of Colonialism in the Americas. As a humanist with specialization in colonial Spanish America, transatlantic studies and contemporary culture of the Caribbean, I wanted to create a course that would make what students perceive as "old stuff" relevant to twenty-first century struggles to maintain cultural integrity in the face of myriad economic and political pressures. Moreover, I wanted students to understand the value of culture through the lens of the humanities, rather than the social sciences; culture as an expression of humanity, community, beauty and resistance, rather than an explicit object of study. I also intended that students would develop the rhetorical skills that the humanities promote: critical reflection of perspectives, argumentative writing and public exposition. Finally, I wanted the Cultural Survival course to present upper-level students with an opportunity to integrate their diverse learning paths and to incorporate their study away experiences into their coursework. Thus, notwithstanding its humanist bent, the class remains quite interdisciplinary in nature and is cross-listed with both Latin American Studies and International Studies. The course fulfills Area I in our major, Origins and Beginnings, which is the required part of the Hispanic Studies curriculum that extends from pre-conquest Native American expression through the early nineteenth century, and it is taught in Spanish.

The primary objective of the course is to trace with students the historical trajectory that connects colonialism with contemporary struggles for cultural survival in the Americas. We use historical texts—primarily from the seventeenth century—testimonial documents, and a range of contemporary critical and secondary texts to explore how and why cultures and languages have been threatened, particularly among peoples of Native American and African descent in selected sites of the Caribbean, Central America, Mexico and North America. We also examine colonial and postcolonial visual arts and verbal and performative expressions that contain strategies of resistance against dominant culture. So that students may come to understand the extent to which the effort to maintain cultural integrity remains an urgent and local concern for many peoples in our historical moment of globalization, a central part of our course entails a semester-long collaboration with a local community organization that values culture as a means to confront social justice issues, whether that be in education, employment, health care, housing or another arena of concern. Along with various critical reflections that students write throughout the semester, the culminating outcome of this course is the project that students develop to advance the objectives of the organization with which they work. Or at least that was the original plan.

Before I taught Cultural Survival for the first time in spring semester of 2011, I worked closely with the Associate Director of Macalester's Civic Engagement Center to identify four community partners with whom students in the class might choose to work. These were neighborhood and community-based organizations with whom Macalester has established partnerships, and the proposed course intended to allow the college to further develop these relationships through cultural commitments. My colleague and I met with representatives from the partner organizations to explain the course content and goals. The objective was to allow pairs or small groups of students to select the organization they would like to partner with, establish an attentive dialogue with the organization to learn about their needs and goals, and then develop in conjunction with community members a project that addresses their efforts towards cultural continuity. Examples of collaborations might be the implementation of an educational program for children that includes activities surrounding the community's language, art, or music, or the development or promotion of an event that heightens local awareness about the community's culture. I was adamant, however, that my students would accommodate the needs of each organization, and that our intention was not to impose a pedagogical agenda. Course readings, presentations and workshops would prepare and guide students in strategies of interaction with community partners, but they were expected to gain important collaborative skills through their direct, hands-on experiences.

One of the first obstacles I faced in trying to accommodate all of my students in community sites for the course was the immense competition for the same sites from other institutions of higher learning in the Twin Cities. Macalester College is fortunate to be located in the urban setting of Saint Paul-Minneapolis, which offers myriad opportunities for local community service. But there are 12 universities and colleges in the Twin Cities area, and as I have mentioned already, civic engagement has become very desirable as a college course component, both from the student and administrative perspectives. (Indeed, the marketing push for civic engagement from college administrations merits close critical examination as we continue to develop our community collaborations.) Whether or not this boom is good for community organizations depends on the perspective of the organization. But while many non-profits could not function without the contributions of student labor, others find their already limited staffs overwhelmed with the need to train a new batch of college students every semester, most of whom will not remain after their coursework ends. I return below to these types of partnering pressures that nonprofits and local organizations currently face. Finally, even though Macalester College maintains respected, long-standing relationships with numerous nonprofits and community organizations, so do many of the other local institutions.

Related to this challenge of demand was the reality that not all of my students could be placed with organizations that were directly related to the geographic focus of our course material. For example, many of our Hispanic Studies majors are most interested in partnering with Hispanic and Latino community organizations, and indeed, a good number of the students in my course were able to work with organizations like El Colegio, Guadalupe Alternative Programs or Centro, all of which focus on youth and education. Other students, however, had preexisting connections to the large Somali or Hmong communities in the Twin Cities and wanted to continue the relationships that they had already established during one or more semesters. At first I was hesitant for students to diverge away from the organizations I had planned for them to work with. But the course offers a solid theoretical framework that prepares students to read society and culture critically. And as I considered our critical readings, which examine neoliberalism and its concomitant forced migrations, international competition for local lands and resources, commodification of cultural expression and imposition of a homogenizing, Anglocentric assimilation (particularly in the insistence on the use of "proper" English), I realized that in all of these communities were experiencing at least some of the pressures on culture that we examined in class. I determined that I would be able to manage the incorporation of other global regions within the purview of our theoretical focus.

Beyond the placement of students at unexpected sites, a few were involved in either fields of study or tasks for the organizations that I had not anticipated. For example, one student whose primary interests were food justice and organic agriculture worked as an intern for a neighborhood organization, Frogtown Farm, that was writing grants and negotiating directly with the city of Saint Paul to secure land where they could a create desperately needed green space for lower income children and set up a community exhibition garden. Another worked as an intake specialist in a mental health clinic in Minneapolis that served a largely Somali clientele. Yet another helped in a food drive to stock the food pantry of Hallie Q. Brown, a nearby African American social and cultural center in a neighborhood called Old Rondo. One of my primary concerns, and certainly that of my students, was to fulfill what each organization most needed at the moment. I weighed the benefits of achieving continuity of collaboration with a community-based organization and those of site experiences more closely aligned with course content. In the end, I felt that sustaining relationships better benefited both the CBO and the student if managed carefully by me. Moreover, in some cases, a site less perfectly suited to my course was also more appropriate to the student's overall academic pursuits, as was the case with the student who worked for Frogtown Farm.

I therefore determined that I would accommodate students who were truly motivated to work at a particular site and help them to locate their service experience within the framework of the course. In the end, while not ideal, it has not been as difficult as I might have imagined to adapt to the challenge of multiple engagement sites and diverse types of student work. In fact, I would even argue that the multiplicity of experiences in a single class has enriched our discussions and brought the lesson home that struggles for cultural survival, after all, are intimately tied up in a web of socioeconomic factors, including education and the imposition of a dominant language (English), as well as other essential concerns such as housing, health care, and transportation. And so, the student at Frogtown Farm became an active grant writer, but also produced a wonderful piece on the neighborhood (which she shared with its representatives) that spoke about the neighborhood's demography, each ethnicity's particular gardening practices and how the new green space would nurture common understanding and better nutrition among children. The student who worked in the mental health clinic wrote a very interesting piece about how Somali traditional medicine and conceptions of mental wellbeing often clashed with common Western medical practices used at the clinic. She noted that the clinic had only begun to be more aware and sensitive to the cultural needs of its immigrant patients. And the student assisting with the food pantry wrote a historical analysis examining how Hallie Q. Brown Center had moved away from its political activity of earlier years had become more attentive to the most urgent needs of the community: food, clothing, childcare, etc. It was a sobering revelation to see such a transformation, as it revealed how desperate our immediate human needs have become in the US.

So, what strategies work to generate an adaptive yet effective pedagogy in a classroom that is student-centered and engaged with community? I have mentioned already the importance of laying a solid foundation of theory and criticism in both the course content and in civic engagement practice, and making clear to students from the start what it is that we are hoping to achieve in our inside and outside of class endeavors. Early in the semester, I ask students to read David Harvey's *A Brief History of Neoliberalism/Breve historia del neoliberalismo* (2005, 2007), Walter Mignolo's *The Idea of Latin America/La idea de América Latina* (1991, 2007), and portions of Néstor García-Canclini's *Culturas híbridas* (1989, 2009). These texts provide students with initial perspectives on how to read colonialism and the neocolonial motivations of neoliberalism in the twenty-first century. Additionally, five times throughout the semester, I give students questions that integrate our textual and document analysis with civic praxis so that they may produce a three-page critical "meditation." I develop the meditation themes according to how the course is evolving organically, and try to leave them sufficiently open-ended to invite various perspectives informed by distinct experiences. Students share their essays with three other students two days before class. Group members, whom I scramble for every meditation, read and comment on each other's writing, and we spend almost an entire class session in our meditation groups tying theory and practice together. The students lead the class because it is their work that has determined the content. This reflective work that I call meditation is key to both developing critical writing skills in Spanish and to integrating the classroom and civic components of learning.

Another of my student-centered objectives for the course has been to create a forum for students to share their study away experiences as relevant to on-campus classroom learning. Many of my students in the class when I most recently taught the course had either just returned from study away or had been back in the US for only six months, and they were still eager to talk about their experiences.

All of those who had studied in Latin America, and even those who had studied away in Spain, had first-hand accounts of communities who were struggling against international mining, forestry, tourism or other economic pressures in order to protect their territories, culture and language. My students now had a meaningful context in which to tell these stories of resistance. They incorporated field research that they had conducted abroad into their meditations and class presentations, and our class discussions were often peppered with examples from their experiences in Peru, Bolivia and Ecuador. I also invited five other students not enrolled in the class to come and present on their honors theses and capstone projects that treated cultural survival themes in Latin America. This exchange with peers gave my students who have not been abroad an opportunity to see how the study away experience teaches students about global concerns, but also connects with their curricular and local engagement endeavors. Some students, after all, will never have the opportunity to study away.

Our work in the Cultural Survival course stresses the commonalities among narratives of cultural struggle, no matter where in the Americas they are occurring. Oftentimes, students in institutions of privilege imagine that problems of inequality and the violence the injustice generates are problems of the developing world. But our focus on local issues and domestic US concerns reveals the unpleasant truth that globalization opens world markets, forces migrations, puts pressure on resources and generates misery *globally*. We see the impacts in our own Twin Cities of Saint Paul and Minneapolis. But in spring 2011, with the help of a Project Pericles Fellowship and support from my institution, I was also able to take several Cultural Survival students to New Orleans over the January term to witness extreme examples of how neocolonial economic and ecological neglect can put cultures at risk in another part of the US. They met with community organizers, scholars and activists, and learned first hand how both urban and wetland communities have developed strategies of resilience. The students who participated in the J-term course gained insights on community organization in the wake of disaster that allowed them to make invaluable contributions to class discussions. My hope is that it also informed their civic labors in local spheres.

Perhaps the greatest challenge that arises from all of the unanticipated variables of the class is helping each student to develop a final project that relates to his or her unique community engagement experience. Because I have decided that students' commitment, passion and academic interests should guide both their community work and research, and because I have intentionally allowed such latitude in their civic engagement experiences, I have also brought a greater workload upon myself as regards their writing. If our course enrollments were larger, I probably would not be able to manage this time and energy demand, because long conversations that involve helping students generate creative insights and critical thinking about their particular engagement are requisite for solid projects. Of course, these consultations involve radically changing gears with each student as we consider the unique aspects of his or her theme, which at times can reach beyond my area of specialization (such as the mental health project did).

I should add that this year, I felt obliged to respond to what some students had stated in their student evaluations from last year, and that was that they were simply too busy to dedicate several hours per week at a CBO site on top of fulfilling the obligations for their campus courses. I am sensitive to the enormous commitments that students have with their studies, jobs and extracurricular activities. And so, I made the civic engagement portion optional this time, requiring that a student who chose not to do a community collaboration would write a 20-page final project instead of a 12 to 15 page project. It was a pleasant surprise for me when only four students out of 16 chose to do the longer research project. And in each of these four cases except one, the longer project was informed by the student's study away experience. So, even these students were able to incorporate into their writing some reflection about important learning experiences with community.

I have spoken quite a bit about the pedagogical flexibility required by the professor in teaching a student-centered, civic engagement course. But I have not made much reference to the students' ability to adapt to the unexpected, and this, in fact should be one of the most important learning goals of such courses, in my view. The Millennials, a term used to describe the generation of North Americans born between 1980–2000, are the students that we currently teach in our classrooms. They are highly connected to information and social networks, and as a general rule—at least from my experience—tend to be proficient at teamwork and collaboration because they are adept at sharing what they know and listening to others' ideas. This ability certainly supports the interactive, student-centered classroom. But there are other defining attributes of this generation that do not. For example, although Millennials have access to an astonishing amount of information, I have found that their understanding of history often does not run very deep. Therefore, we have significant ground to cover before they can grasp the full historical weight of social justice issues like racial discrimination, voting access or equality in education. Millennial students are often creative and innovative, but at the same time, they want to have explicit instructions on what they are supposed to know, and they display anxiety when a clear, precise structure is not in place. Whereas I used to be able to give students total freedom to determine research and writing topics, I find with Millennials that they become overwhelmed by infinite possibility, and for every assignment, they want detailed instructions that explain exactly what is expected of them: how the assignment should be organized, what specific information it should contain and precisely how it will be evaluated. A colleague has called this anxiety the result of Millennials being "No Child Left Behinded to death." Needless to say, an overly formulaic assignment can produce rather formulaic, and at times boring, results.

Related to this quasi-prescribed learning tendency, our contemporary students, while adept at working in groups, also periodically state in course evaluations their preference that course knowledge come from the professor, that is, someone they consider to be a reliable authority figure. They claim that they find group work and student presentations to be less informative, and thus less beneficial to them. This may in fact stem from a desire to understand the beliefs, values and positions of the person who controls their grade in order to best meet those expectations. I also suspect that they are unaware that important learning is taking place even in the process of clarifying muddled thinking. I am sensitive to substantiated comments from students on course evaluations, and I respond to them to the best of my ability. I seek balance between what I know to be sound pedagogical practice and the preferences that my students share. All of that being said, I believe that Millennial students have much to gain from a learning environment that is rigorous and organized, but at the same time open to the rich possibilities of learning as a collaborative process instead of learning as a commodity of privilege passed between one "who knows" and one who does not. It is for this reason that I maintain student-led discussions, group work and class presentations as key components of my classroom teaching. It is also why I want my students to work in tandem with people outside of the college where they will recognize that valuable knowledge is produced in other arenas as well.

Perhaps it is easy to see how college students can be enriched by their civic engagement learning experiences. But as we contemplate the ideal partnerships that we seek to create between our students and myriad community organizations, nonprofits and schools, we must take a moment to ask ourselves honestly if community partners reap equally meaningful benefits from collaborations with college students and the institutions they represent. There is no single or simple answer to such a question, as each engagement scenario is distinct. Certainly, the most valuable civic engagement relationships allow all parties to feel as if the collaboration has advanced their mission or assisted their members, residents, clients or students in some significant way. Yet, we know that the power differential and the consumptive capacity of the town-gown relationship can leave organizations feeling depleted rather than enhanced. Underscoring this sentiment, Seitu Jones, a public artist and resident of Frogtown in Saint Paul, Minnesota, stated in the keynote address of the 2011 Imagining America conference held in the Twin Cities that he feels that his neighborhood has been repeatedly explored, interrogated, dissected, and put under a microscope by university civic engagement projects.[2] He assured the audience that Frogtown residents are grateful for the care, but that sometimes the attention has felt excessive.

One must interrogate, too, the type of work that students perform for local organizations. In civic engagement collaborations, do all parties envision the project and its outcomes together? Is the relationship truly reciprocal? Is the collaboration transformative for society? Does it serve the common, public good?

---

2   Frogtown is a historic working class neighborhood just west of downtown Saint Paul, and close to several institutions of higher learning. Imagining America was formally launched at a 1999 White House Conference initiated by the White House Millennium Council, the University of Michigan, and the Woodrow Wilson National Fellowship Foundation. A consortium of 90 colleges and universities, and their partners, IA emphasizes the possibilities of humanities, arts, and design in knowledge-generating initiatives (http://imaginingamerica.org/).

The spectrum of possibility and utility is wide. For example, the impact of regular one-on-one tutoring between a college student and a child in elementary school may be more immediately apparent than that of a student who answers calls from undocumented immigrants seeking health services or legal assistance. The students and neighbors who collaborate to imagine, design and construct a playground for children who have no play space take community engagement a step further. And yet, all of these roles are in fact desperately needed in our communities.

In recent years, the most significant impact on the ability of nonprofits to meet community needs in the US has been the economic crisis. As the 2008 housing crisis and subsequent market crash evolved into a recession that we continue to endure in 2013, many sources of funding for nonprofits have disappeared. Deep government cuts in public spending have combined with a decline in giving from the most generous sector of our country (the middle class that is also feeling squeezed) to create a dearth of resources precisely at a time when our populace is in most need.[3] Journals such as *Nonprofit Quarterly*, *Philanthropy Journal*, and the *Chronicle of Philanthropy* have all documented over the past few years the negative impact of the funding crunch on nonprofits that provide essential resources and programming to underserved populations in the US. Alarmingly, as government funding sources have diminished, corporate giving to nonprofits has increased, which leads one to question what the donor expectations might be for acceptance of those funds. We should also contemplate what the structural and moral consequences might be for long term reliance on such donations.[4]

University and college students in civic engagement projects, therefore, are often contributing their time and energy to organizations that have been stretched to their financial limits and may have even been obliged to cut permanent staff positions. I mentioned above that nonprofits can be overburdened with the ongoing cycle of training college interns and volunteers who are present only for the duration of a service learning project—generally a semester—and then depart. But the difficult truth is that many organizations have come to rely on students as unpaid staff in these financially difficult times, and it is sometimes in their best interest to try to retain student volunteers even after the civic engagement project or internship has ended. It is in part for this reason that sustained and consistent relationships between colleges and local organizations are so important: above all, this continuity in turn helps community organizations to create stability for populations who rely on their services.

As our students collaborate with nonprofits, they need to become aware of the circumstances that make their labor within these organizations so desperately

---

3   See "America's Generosity Divide" in the *Chronicle of Philanthropy* (August 19, 2012), http://philanthropy.com/article/America-s-Generosity-Divide/133775/.

4   See "Giving USA 2013: Giving Coming Back Slowly and Different After Recession" in *Nonprofit Quarterly* (June 18, 2013), http://www.nonprofitquarterly.org/philanthropy/22476-giving-usa-2013-giving-coming-back-slowly-and-different-after-recession.html.

needed. They need to realize that they are fulfilling roles that are now crucial in North American society, and they should critically interrogate this scenario as they work to remedy it. Is it desirable that essential community services and education for the general population should rely so heavily on temporary, shifting student and volunteer labor? Or should we demand more stable and consistent funding for these basic services from government sources? Furthermore, is it morally sound that our bright and creative youth, already burdened with enormous student debt and myriad social, environmental and political woes, should shoulder the responsibility of our national neglect free of charge? Clearly, yet another learning outcome of civic collaboration within the community is that students can begin to contemplate what type of future they want for their country.

The decentering of the traditional college classroom is as important to nourishing democracy as are collaborations between higher education and local communities, when properly conceived. I believe that students' ability to rely on each other as valuable producers of knowledge and to rely on each other as teacher-learners parallels the ideal of their role as responsible citizens who can negotiate difference and meet others on equitable grounds of mutual respect and trust. In both of these processes, there are challenges of discomfort, uneasiness and uncertainty. But students grow in myriad ways as responsible members of society from learning how to peaceably navigate an uncomfortable scenario, particularly in a volatile political landscape. They also become more creative and confident learners when they can integrate unexpected events into their work. This is precisely why both professor and student should embrace the spontaneity of classroom and civic collaborations.

As I mentioned above, Walter Mignolo's *The Idea of Latin America* forms part of our critical framework in the Cultural Survival course. In his thoughts about decolonization, Mignolo talks about the importance of inhabiting what he calls the "colonial wound" in order to recognize and overcome the violence of the colonial past and work for real justice in our neocolonial present. When one is willing to confront history honestly and allow oneself to enter into a new epistemology that differs from the modernist/colonialist paradigm, true interculturality can take place, that is, an exchange of ideas that rejects cultural superiority and dominance. As my students examine their own positionality *vis-à-vis* some of the communities with whom they work, they become uncomfortable, and rightly so. They wonder what role they can possibly play in a community that is not their own. They wonder what responsibility they have in replicating oppression. And I encourage this questioning and discomfort. I ask them to try to inhabit the colonial wound, as Mignolo suggests. One of our meditations, in fact, asks them to examine their own role, their institution's role, and even the role of the organization with which they collaborate in processes of decolonization or neocolonization. But I also remind them that without solidarity across society, we will get nowhere in our efforts to remedy injustice. They often need to be coaxed out of what is called "analysis paralysis:" over-thinking or over-analysis that leads to inaction. Students also often fall prey to the helpless feeling that critical theory can produce in a student who wants to effect change in the world but who is convinced that s/he has no right to act due to class or race privilege. It is too easy for any one of us to walk away, for whatever reason, and say "There is nothing I can do here." Instead, I tell them, "You cannot stand still. We are in too urgent of a historical moment. Be willing to make mistakes. Be willing to listen and learn from those who know more than you. Stay involved, be critical, keep moving."

A good example of this discomfort occurs when students who work at immigrant centers help teach English to immigrants. They feel that they are contributing to an imposition of dominant culture onto a population that must assimilate or perish in the US. In short, they are correct, and I do not have any easy answers to these concerns. We have learned about the circumstances that push people away from their homes and into new lands, and these are not innocuous forces. And so we have long debates about the pitfalls and the pragmatic benefits of teaching or learning English. In the end, coming to a final conclusion matters less than the process of critically analyzing the many factors that put a population's culture or language at risk and continuing to work in solidarity with local communities that are facing similar struggles.

In returning to the destabilization of the classroom mentioned at the beginning of this chapter, one can perhaps see now how pedagogies that promote democracy may seem less orderly or structured than the traditional classroom. Our plans laid out for student engagement often evolve in ways we could not have anticipated. But by folding an ability to navigate the unexpected into the course as a learning objective that promotes problem-solving, negotiation of difference, and the birth of new ideas, students can learn to be at ease with the discomfort that unscripted collaborations with community often entail. In our efforts to cultivate an ethical relationship to society among students and promote their sense of the public good, we need to transform our institutions in ways that foment democracy both inside and outside of the classroom. And as the process becomes slow and messy and even frustrating at times, both professor and students need to learn to recognize the freedom—and enormous creative, innovative potential—of rolling with the punches, and of making meaning out of events unforeseen.

References

Coalition for Civic Engagement and Leadership. 2005. *Working Definition of Civic Engagement*. University of Maryland. http://www.nl.edu/cec/upload/Working-Definition-of-Civic-Engagement.pdf.

Cronon, William. 1998. "Only Connect …" The Goals of a Liberal Education. *The American Scholar*, 73–80.

Dewey, John. 1916. *Democracy and Education*. New York and London: Free Press.

García Canclini, Néstor. (1989) 2009. *Culturas híbridas: estrategias para entrar y salir de la modernidad*. Mexico: Debolsillo (Random House Mondadori).

Harvey, David. 2005. *A Brief History of Neoliberalism*. New York: Oxford UP.

Harvey, David. 2007. *Breve historia del neoliberalismo.* Madrid, Spain: Ediciones Akal.

Jacoby, Barbara (ed). 2009. *Civic Engagement in Higher Education: Concepts and Practices.* San Francisco, California: Jossy-Bass.

Jay, Gregory. 2009. What (Public) Good are the (Engaged) Humanities? Publications of Imagining America: Artists and Scholars in Public Life. http://imaginingamerica.org/publications/reports-essays/.

Jones, Leo. 2007. *The Student-Centered Classroom.* Cambridge: Cambridge UP.

Mignolo, Walter. 2007. *La idea de América latina: la herida colonial y la opción Decolonial.* Buenos Aires: Gedisa Editorial.

Mignolo, Walter. 1991. *The Idea of Latin America.* New York: Wiley-Blackwell Manifestos.

Strait, Jean R. and Marybeth Lima (eds.) 2009. *The Future of Service-Learning: New Solutions for Sustaining and Improving Practice.* Virginia: Stylus Publishing.

Sullivan, William M. 2005. *Work and Integrity: The Crisis and Promise of Professionalism in America.* Second edition. San Francisco, California: Jossy-Bass.

Sullivan, William M. and Matthew S. Rosin (eds). 2008. *A New Agenda for Higher Education: Shaping a Life of the Mind for Practice.* San Francisco, California: Jossy-Bass.

Chapter 4

Community Enrollment:
Colleges and the Fault Lines Between
Academic and Civic Engagement

Stephen Tremaine

How can our colleges and universities become symbols of civic democracy when our own faculty and students question our commitment to true democracy and civic commitment embodied in concepts of diversity? What happens when the rhetoric of civic engagement smacks into the realities of the current limitations of access and fundamental retreat from concepts of inclusiveness, whatever the root causes?

(Sanchez 2005, 11)

I like to think that, occasionally, a New Englander riding the Carondelet streetcar through downtown New Orleans looks upward at a faded concrete high school building and is perplexed: a bright red sign juts out from the third floor, proclaiming a campus of Bard College.

On this campus, on any given school day, a truly unusual collection of Bard College undergraduates can be found in New Orleans. They are unusual on several counts: they are 16 and 17 years old; they spend much of their lives in New Orleans' public high schools; and they are, for people of any age, uncommonly committed to the values of open inquiry, dialogue, and curiosity that define a liberal education. The Bard Early College in New Orleans (BECNO) is a satellite campus of Bard College (of New York State's Hudson Valley) embedded within New Orleans's public high school system. Founded in 2008 in response to a call from the Louisiana Department of Education for more intellectually engaging classrooms, BECNO enables younger students to begin undergraduate coursework in the liberal arts during the last two years of high school—years that, for many students, are defined by intellectual frustration, stultification, and remediation.

BECNO enrolls over 130 11th and 12th graders every year. These students start their days as high schoolers and end them as Bard College undergraduates, completing basic high school coursework at a traditional public school and riding a yellow school bus to Bard's downtown campus for the second half of the school day.

Just as BECNO students are atypical undergraduates, Bard's engagement in New Orleans is unconventional in the national landscape of postsecondary civic engagement. BECNO is not described by any of the familiar categories of civic engagement: service learning, engaged scholarship, community-based research. In higher education, we are often accustomed to thinking through the dynamics of civic engagement in terms of the impact that engaged scholarship has on our students, on ourselves, and on the communities that we collaborate with. In New Orleans, Bard College takes an unusual step toward unraveling these distinctions: rather than student, university, and community, Bard aims to collapse the three; rather than community service, Bard's ambition is toward community enrollment.

BECNO can perhaps be of greatest interest to the conversation around academic civic engagement less as a model for replication than as a point of departure for a different line of thought about universities and underserved communities. In its short and clumsy history as an oddity in New Orleans' public school landscape, BECNO has shed light on critical questions about what we represent, who we serve, and what resources we aim to mobilize. What are the implications of this particular form of university civic engagement—what I am calling here, for lack of a better term, community enrollment—and what does it reveal about the place of colleges acting in the public interest? Civic engagement has been critiqued for evading the forms of intellectual accountability that have strengthened other elements of the academy, including curriculum and faculty (Harward and Finley 2012). In aiming to ensure that civic engagement is something for which colleges are intellectually accountable, we should question old models—and explore new ones. In a small way, I believe that the notion of community enrollment can help us to do both.

First it will be useful to have a better picture of our point of departure, BECNO, and of what brought it to life. Bard has a substantial history of working to extend the transformative potential of college education in the liberal arts and sciences to bright, often under-engaged younger students. Bard runs a network of public high schools (two in New York City, one in Newark, New Jersey) that enable students to complete two years of a Bard education during high school. The faculty of these unique schools hold terminal degrees in their fields; they are publishing and are actively engaged in scholarship. They model a depth of intellectual and critical engagement often lacking in the American public high school.

Students on these campuses are often bright young people who have found high school slow, punitive, and reductive. And, indeed, public high schools—burdened by the pressures of high-stakes testing, vocationalism, and remediation—have been pushed farther from the forms of critical thought and intellectual engagement that define many of the liberal arts colleges they hope to prepare students for. This is particularly true in high-poverty urban school districts, where students face the most daunting chasm between secondary and postsecondary education—and, often, the weakest bridge on which to cross.

As an undergraduate at Bard, I saw my classmates—all of them fiercely smart, critical, and committed to making the opportunities they'd had more widely available—aspiring to become teachers in a moment when public education was, predominantly and often unambiguously, anti-intellectual. I badly wanted to see a Bard-run high school in my hometown, New Orleans, particularly after a hurricane blew apart the school system that had been. When I graduated, Bard's President, Leon Botstein, gave me a job and an invitation to build such a campus. BECNO, as it exists today, is all of eight rooms with chalkboards and seminar tables. Throughout the last two years of high school, our students complete the first year of an undergraduate education in the liberal arts. Our faculty teach concurrently at local universities, and many have taught farther afield; most have sought opportunities throughout their careers to engage students making the transition to college.

It would be misleading to think of early college simply as accelerated high school. Programs like BECNO are not about asking students to study the same material in a more advanced textbook, but rather about helping them learn new ways of thinking critically, expressing intellectual curiosity, and engaging in critical analysis. In this sense, the early college seminar is less about the questions that students are answering than the questions they're asking.

The Bard Seminar in New Orleans—modeled on the Bard First-Year Seminar on Bard's home campus—introduces students to the college classroom through a question central to study in the liberal arts and sciences: what does it mean to be human? How have scientists and thinkers across traditions attempted to articulate a common understanding of humanity? Students read Friedrich Nietzsche and James Baldwin, Jamaica Kincaid, Judith Butler, Carl Sagan, and Rousseau, a trajectory across much of the modern intellectual tradition.

BECNO students come from every zip code in New Orleans. They are concurrently enrolled in over 85 percent of the open admissions, 9th to 12th grade high schools in the city. BECNO students are chosen by audition; they read an essay in preparation for a trial class session with 14 fellow applicants, and they are evaluated on a single measure: when faced with a difficult point in the text—an unresolvable debate, a problem that resists reduction to a simple right or wrong answer—do they, at that point, get excited and engaged by the text? Or do they disengage, revert to platitudes, resist deeper thinking? The applicants who, over the course of a 90-minute, seminar-style conversation, seem to engage when faced with ambiguity, we accept. We know nothing at this point about their GPA, their PSATs, or any other bits of the trail of numbers that they've accumulated throughout their lives in classrooms—the numbers that most often define them in the college admissions process. While faulty, subjective, and probably biased towards the more outgoing students, this is our best attempt at addressing an essential and elusive problem: distinguishing between the quality of a young person's preparation and the quality of his or her mind. BECNO students have, by several measures, done well. To date, 98 percent of BECNO students have matriculated to full-time undergraduate enrollment. 97 percent of alumni since 2011 remain enrolled in college.

Now, five years into growing, tinkering with, and worrying over this effort in New Orleans, it seems inadequate to refer to it as a civic engagement project. BECNO is a *bona fide* campus of Bard College. Its students are part-time Bard undergraduates. We enroll—quite competitively—young people who are expected to meet the standard of inquiry of a traditional Bard classroom. In this sense, it would appear to be simply a franchise of the College. And yet BECNO represents something other than a conventional outcropping of the institution. Like other civic engagement projects, it represents a response to a perceived social problem (lack of meaningful bridges to postsecondary education) in a specific time and place (post-Katrina New Orleans) on the part of a college. As with other civic engagement projects, we have taken extraordinary measures (endless fundraising) to make this opportunity free of cost for a group that we, at Bard, have deemed underserved (bright, disengaged adolescents in New Orleans' public high schools).

BECNO is caught between these two categories. In this sense, early college programming illustrates an intersection between Bard's identity as an academic community and as a civically engaged institution. Those two aspects of a college's work—academics and civic engagement—are traditionally discussed separately. Their neat separation is neither accurate nor salutary. Bard's work in early college education provides a chance to consider how, at Bard, they are fundamentally congruent. BECNO is not exceptional in its resistance to the most readily available categories of postsecondary engagement. Many worthy-but-complicated efforts fall between the categories I am familiar with. The insufficiency of these categories deserves our attention.

I suggest that much institutional engagement has centered on two strategies: student-centered projects and faculty/staff-designed service learning. Attempting to fit community enrollment projects such as BECNO into these categories reveals important questions about the culture of service and civic engagement in higher education, particularly the points at which its claims to intellectual accountability may be open to critique.

Student-centered Projects and Institutional Engagement

A reversal of one of the orthodoxies of civic engagement is at the heart of BECNO. Civic engagement at colleges is often introduced in the form of a catalogue of opportunities for traditional undergraduates to contribute to communities of need; the invitation is for the student to act, and for the institution to facilitate. In the case of BECNO, the reverse was true: emphasis was on the institution acting and, in so doing, creating new student communities. What does it mean, then, for an institution—not simply its students—to be civically engaged?

When, in the semesters following Hurricane Katrina, Bard undergraduates asked "What is most needed?" the result was a pantomime of real home renovations; students at Bard (myself foremost among them) sought to help out as amateur carpenters, electricians, and contractors. We were notably bad at it. We within the student body determined, ultimately, that our efforts at service were ineffectual and, too often, one-sided: while roofs and floors and walls unquestionably needed to be fixed, we were also thinking about what that form of involvement led to for Bard. There was clearly much more that we as an institution could do, and perhaps things that were badly needed that we could do uniquely well. These questions led to a conversation between students and administrators that, in short order, charted the transition from a student endeavor to an institutional initiative. When we asked not "What is most needed?" but rather "What do we do best?" the result was BECNO, an effort that Bard remains deeply invested in and responsible towards, an effort that reflects what we do well as an institution.

BECNO makes sense within the culture of Bard; it is an expression of the college's unique values and its particular history of engagement. Bard's programming has long focused on extending the academic resources of the college into non-traditional and socially, politically, and economically challenged settings. A talent for bringing the liberal arts and sciences to social institutions (high schools, prisons) and communities in which they're often least accessible is something that defines Bard.

These programs include the Bard Prison Initiative, which enrolls incarcerated men and women in New York State in a Bard liberal arts curriculum. Over 200 students in that program—the largest higher education program in the New York State correctional system and the largest college-in-prisons program in the country—work toward Associate's and Bachelor's degrees from Bard.

There are many other examples. Bard's Clemente Course in the Humanities offers an intensive introduction to the liberal arts for adult students, many of whom are seeking to enroll in college after years outside of classrooms. Bard undergraduates have established a series of writing workshops in Palestine's West Bank.

Community enrollment is, in this sense, part of Bard's core identity as an institution. Civic engagement projects—and the communities they aim to benefit—are well served when colleges and universities begin with an understanding of their unique strengths and, from that point, ask where and for whom that resource could most powerfully be brought to bear. The risk, otherwise, is that an institution's good intentions will lead it to foster civic engagement projects that are undertaken carelessly, built upon weakly or not at all, and not clearly accountable to a standard that the college values. Bard's involvement in New Orleans, initially, focused on the problem (the desolation immediately following Katrina) and worked backwards from there to find a relevant resource at our disposal (our well-meaning but inadequate rebuilding efforts). The College has found far more success in doing the reverse: starting with an understanding of what, specifically, we do well, what sets us apart and defines us, and seeking out opportunities to make that talent of greater social use.[1] If anything about BECNO merits widespread replication, it is this approach.

---

1   This is another point at which a predominant way of speaking about colleges, service, and the public good may prove counterproductive. An emphasis on "best practices" too often takes focus away from the particulars of a place, a community, and a college, and instead suggests that civic engagement projects can always be successfully uprooted and replicated. Perhaps the best practice for such projects is close and thoughtful attention to an institution's strengths, intentions, and the needs of its partners.

Ultimately, Bard acted to facilitate student initiative (myself and my classmates arguing for a more meaningful, education-focused effort, in place of our disappointing attempts). But it did so by, effectively, handing students the keys to the institution: the president of the College empowered me to develop a new branch of the institution. This is an extraordinary step, and not without precedent at Bard. The Bard Prison Initiative (BPI) was founded at Bard by a student. BPI is the result of a student's inspiration to extend what was most transformative about the education he'd received at Bard to men and women with few opportunities to take on serious intellectual work. Bard's Trustee Leader Scholar (TLS) program serves as a bridge for dialogue between students and the College leadership. TLS vigorously backs student initiative. Students in this program design and lead projects in the West Bank, in Nepal, in Nicaragua, and throughout Bard's region of New York. The promise it offers students is of an institution that will meet their ambitions for social change—an institution that is liable to call their bluffs and inspire them to take risks.

Blurring the lines between student initiative and institutional action has thus been a powerful way for Bard to challenge conservative tendencies in the college, to broaden the institution's imagination, and to exploit—to good effect—the potent naivety of undergraduates looking to take on ambitious issues. It has also contributed significantly to the accountability of this work. Rather than a peripheral project, catching the administration's eye only when something may be press-worthy, these efforts have required, categorically, the direct involvement and collaboration of all levels of administration, as well as faculty and students. More than isolated initiatives, they are lasting extensions of the campus into new communities.

The risk here is, obviously, that a student-turned-administrator will act in ways that, whether negligent, incompetent, or inappropriate, embarrass the institution and mistreat the communities that the college had set out to provide for. There is likely no remedy for this. In my case, the figures who were best positioned to help me were in the foundation, public education policy, and higher education landscapes. These arenas are typically slow to change and deeply dependent on networks and personal pedigrees. My joining this world as a 22-year old college graduate was no less strange for me than it was for the college presidents, superintendents, program officers, and elected officials into whose offices I entered—new to the business of shaving regularly, without business cards or the traditional weaponry with which one advances one's cause, and learning haltingly from online videos how exactly to tie a necktie.

These risks are counterbalanced, though, by a sense of accountability, investment, and engagement shared across the institution. Too often, student engagement at colleges and universities is defined by obligation, often quite literally: increasingly, colleges and universities have taken to mandating community service among their undergraduates. (That these decisive new investments in public service have coincided with a tremendous shift of resources away from the very academic departments that have given us discourses for reckoning with the public good—literature, philosophy, anthropology, and the other fields that have taken the brunt of recent budget cuts in major research institutions—does not make for a happy balance [Bard and Turner 2013; Palacios et al. 2013]). In the worst cases, students made to act involuntarily in the public interest are encouraged by their universities to view the social good as a mandate, in the same passionless category as taxation and dental hygiene. The university becomes the place in which students find their intellectual lives estranged from public service, rather than the place in which they find the rich and shared space between their intellectual passions and the public interest. When those students, on the other hand, see their institution as actively and dynamically engaged in the same forms of social inquiry, critique, and engagement that students often model, they are given reason to think more ambitiously and more accountably about civic engagement. When the institution takes real initiative to engage students in dialogue around civic engagement, it becomes more than just a source of predetermined options for students to "serve," and its students become more than just passive participants in others' projects.

Service Learning and Learning as a Service

Another predominant form of civic engagement at many higher education institutions is service learning. Campus Compact, a national organization that works to foster civic engagement at colleges and universities, defines service learning for their network: "Service-learning incorporates community work into the curriculum, giving students real-world learning experiences that enhance their academic learning while providing a tangible benefit for the community."[2] While this work is often of real social and humanitarian value, a troubling implication is bound up in the name, complicating the intellectual accountability of this form of engagement: in embracing these notions of service learning, are we suggesting that learning is not, itself, a service? While the seminar and the "service," the "academic learning" and the "tangible benefit," do not explicitly compete with each other, we can't ignore the implicit sense fostered on some campuses that students should perform acts of community service to pay down the moral and social debt incurred by all of that self-indulgent learning. The very definition of service learning that Campus Compact provides encloses the suggestion that "academic learning" is not, itself, a "tangible benefit for the community." The ways of framing education and society to which this leads reflect a fraught vision of colleges as divided in purpose between general education and social responsibility.

In its most careless forms, we allow service learning to suggest that the traditional undergraduate seminar—dialogic, demanding, committed to thoughtful attention to difference and the sussing out of historical, cultural, and scientific significance—is somehow not a socially and civically vital service. Civic engagement at Bard has taken shape against these themes; indeed, it is worth noting that Bard's most socially transformative initiatives—from reshaping prison education to rethinking

---

2   http://www.compact.org/initiatives/service-learning/.

the American high school—are extensions of the classroom. The early colleges are nothing more and nothing less than Bard campuses.

In post-Katrina New Orleans, the classroom opportunities for adolescents from devastated communities in New Orleans were largely premised on narrow vocational training or test-driven instruction. For young people whose lives and communities had grown exponentially more chaotic and more impervious to simple solutions, something urgent and timely was lost when the classroom aimed to reduce big problems and big questions to binary, testable, and verifiable answers. Bard argued that the most meaningful service we could offer would be to make a liberal education—one rooted in a tradition of free and open inquiry, of critique and social analysis—available to the young people who would, in a few years, inherit a city that needed literally to be put back together. The talent that the city needed, it seemed, was that of questioning old truths, of risking new ideas. This is a talent that colleges like Bard could credibly claim to foster and support; this is an essential project no matter where the campus or who the student.

When we offer learning as a service, we are not fixing broken infrastructure or proposing innovative policies to local governments. We are bringing the curricula, teaching, and conversations that define our institutions to a wider audience. Bard's Early College in New Orleans, rather than making well-meaning scholarly contributions to those less fortunate—or giving out day-passes to academia—aims to foster a truly democratic and diverse academic community, separate from each other and Bard's "traditional" campus in New York's Hudson Valley only by geography. It is precisely in this sense that Bard's approach is best understood not as community *engagement* but rather as community *enrollment*.

Community enrollment positions the members of underserved communities alongside a college's traditional student groups in a strange and unique way. Both groups are given a legitimate claim to the college classroom. Both are formally enrolled. The non-traditional underrepresented students are endowed—to the greatest extent allowed by the circumstances—with the rights, institutional capital, and access to resources that the traditional students enjoy. They are our students and not our charitable experiments. As such, they can expect the college to be accountable to their wellbeing in the classroom, to their learning, and to their experience in the college community, however largely writ.

This ambition is, of course, never perfectly met. We fall short, in New Orleans, in convincing our students in earnest of their place at the heart of a prestigious institution with real cultural capital. We bestow on them the credits, credibility, and pageantry of the college, but their campus in New Orleans lacks some of the basic resources that define the traditional campus: full-time faculty, a real library, and gorgeous science labs. At BECNO, in lieu of teaching laboratories, we have a stretch of parking lot on which we blow things up during physics experiments.

The rhetoric of academia makes this disconnect more apparent; what does it mean to inhabit the *satellite* campus? What qualifies the center? This language presupposes a relationship of oversight and bureaucratic parentage between the two, the traditional campus pulling fledgling bodies into its orbit. At the heart of this shortcoming is an essential irony: while we may often fail to persuade BECNO students that they are more than peripheral to the college, we perhaps also fail to remember that they are precisely what sets that college apart. Any college with sufficient resources can hire Nobelists, churn out expensive degrees, lure big men to play ball. But what our younger students in New Orleans have accomplished—entering the Bard classroom from the most chaotic school system in the country, bearing the weight of a broken city, meeting the standards of the institution in a dramatically different setting than the more familiar embrace of an elite, four-year campus—is something that truly distinguishes Bard. The same is true of Bard's other community enrollment efforts.

In these senses—fostering institutional engagement and not just student initiative, viewing the classroom as a meaningful resource in and of itself, moving beyond the bounds of the traditional campus to find ways to meet new student communities where they're at—BECNO acts less in the spirit of Bard's fellow private colleges of the liberal arts and sciences than of community colleges. It is the core business of community colleges to find ways to bring higher education—as an engine of social mobility—to all communities, including those in which students face daunting obstacles to college. What is, for community colleges, a business model is often categorized, in elite four-year colleges, as civic engagement.

This all begs a return to an obvious question: what's the difference between "civic engagement" and doing what we do as a college? Community enrollment projects raise this question in a particularly stark light. Imagine it: there is a traditional campus, and there is an affiliated satellite campus built for students who are not well represented in higher education. The first is a college; the second is an endeavor in civic engagement. BECNO aims to extend the quality and rigor of the education on offer at the main campus to the students at the satellite. *The project is valuable only insofar as it can successfully match the quality of that education in a less traditional setting.* Are one set of young people our students and the others our projects? As George Sanchez quotes of W.E.B. DuBois, "Being a problem is a strange experience."

Imagining America, based at Syracuse University, is one of American higher education's most active, forward-thinking, and important centers of conversation regarding the civic mission of colleges and universities. IA advances a theory of "institutional citizenship," urging colleges to assume a wider range of responsibilities towards underrepresented communities. The goals of this approach are stated as such:

> Consistent with this vision, [institutional citizenship] uses collaborative inquiry to advance three linked goals: (1) increasing access, success and full participation in higher education for underrepresented groups and communities, (2) building higher education's capacity to address urgent challenges facing these

communities through public engagement, and (3) prompting the institutional re-imagination needed to facilitate the achievement of these goals.[3]

These are vital goals, and IA is well positioned to advance them. It is worth noting the distinction drawn between "increasing access, success, and full participation in higher education" and strengthening "higher education's capacity to address urgent challenges facing these communities through public engagement." Is increasing access to higher education not itself a means of addressing urgent challenges facing "underrepresented groups and communities"?

George Sanchez (2005) challenges the perceived distinction between the two, in his important lecture (to an annual Imagining America meeting) "Crossing Figueroa: the Tangled Web of Diversity and Democracy." Sanchez writes of the "seeming inconsistency of the wide-spread growth of service learning and community engagement at universities across the nation and the rapid decline in programs and commitments to make our own university communities more inclusive and diverse." He argues that positioning the uncertain line between civic engagement and the core business of a college is more than a rhetorical debate: "on the resolution of this inconsistency hangs the role of the university of the twenty-first century as a democratic institution, one that either is able to fulfill its rhetoric concerning civic responsibility, or one that is judged by the communities in which we reside to be full of empty promises and selfish motives (6)."

Sanchez draws our attention, again, to the perceived disconnect between a college's academic identity and its civic engagement. While he doesn't cite them in discussing "programs and commitments to make our own university communities more inclusive and diverse," I would argue that community enrollment projects, at their best, can be uniquely powerful toward this end. They can narrow the artificial divide between classroom and community purpose.

We are accustomed to thinking of civic engagement as a set of projects, qualities, and goals that emerge from a thriving campus. Sanchez asks us to reverse that assumption; rather than secondary to or succeeding the project of traditional undergraduate education, civic engagement is the foundation from which that traditional education gains meaning—a college community that enacts democratic values, that takes extraordinary steps to ensure that the education it offers impacts communities who can gain tremendously from it, is one that cannot be accused of making empty promises. Indeed, for Sanchez, such a campus is the only kind that has made good on its most basic promise.

---

3   (http://imaginingamerica.org/research/full-participation/request-for-proposals/).

References

Barr, A. and S.E. Turner. 2013. Expanding Enrollments and Contracting State Budgets The Effect of the Great Recession on Higher Education. *The ANNALS of the American Academy of Political and Social Science*, 650(1): 168–93.

Harward, D.W. and A.P. Finley, eds. 2012. *Transforming undergraduate education: Theory that compels and practices that succeed*. Rowman and Littlefield.

Palacios, V., IU. Johnson, and M. Leachman. 2013. Recent deep state higher education cuts may harm students and the economy for years to come. Center on Budget and Policy Priorities. http://www.cbpp.org/files/3-19-13sfp.pdf

Sanchez, George. 2005. Crossing Figueroa: The Tangled Web of Diversity and Democracy. *Forseeable Futures, Working Papers from Imagining America*, 4.

Part II

How We Engage:
Modes of Participation, From Digital
Social Media to Radical Democracy

Amy E. Lesen

A promising development is the renewed attention being paid to participation and transparency ... Although these efforts are admirable, formal participatory opportunities cannot by themselves ensure the representative and democratic governance of science. There are, to start with, practical problems. People may not possess enough specialized knowledge and material resources to take advantage of formal procedures. Participation may occur too late to identify alternatives to dominant or default options; some processes ... may be too *ad hoc* or issue specific to exercise sustained influence. More problematic is the fact that even timely participation does not necessarily improve decision-making ... [participation] should be treated as a standard operating procedure of democracy, but its aims must be considered as carefully as its mechanisms.

The above statements by the renowned scholar of science and technology studies, Sheila Jasanoff (177, 182),[1] could easily be the introduction to Janice Cumberbatch's piece in this volume, in which she exhaustively investigates the factors that create success in participation. In the process of doing so, Cumberbatch gives us a precious gift: a framework and a set of checklists that one can use when planning, launching, carrying out, and evaluating a participatory project. She developed this framework via in-depth focus groups and surveys of longtime practitioners of participatory work, and her own years of participatory research in the Caribbean. We come away with a rich appreciation of the important considerations for doing efficacious participatory research, including the types of projects and situations that lend themselves well to participation and, as Jasanoff implies above, the ones that do not. Her work is inordinately valuable for one engaging in participatory research any stage of her career, from the newly-minted to the longtime practitioner.

I thought Cumberbatch's piece the perfect one to begin the more praxis-oriented second part of this book, as she gives us a full vocabulary for participation and a firm basis for discussing modes of engagement in depth, as Kristina Peterson

---

1   Jasanoff, Sheila. 2012. *Science and Public Reason*. London and New York: Routledge.

does in her chapter. I met Peterson at the 2010 symposium I organized in New Orleans, and it became quickly clear that, for those who desire to achieve full parity in the participatory process, where community members are equal partners and owners of the work and are involved intimately at all stages, Peterson's work is a true model. Having got to know Peterson well since 2010 as well as some of the communities she works in, I can attest that she not only collaborates with those communities, but she has come to be a part of them. On the continuum of engagement I discussed in the preface, Peterson exists on the very edge of the most participatory and engaged, and this approach—as will become clear when reading her chapter—will not appeal to every scholar, and it requires deep commitment and understanding of the process.

Peterson's example of the Community Declaration of Principles, her discussion of the power dynamics between community members and scholars and agency workers, and the stories she tells about the empowerment of her community collaborators are substantial food-for-thought for anyone doing or considering civically engaged scholarship. That is evidenced by the transcript of a conversation I had with Peterson and one of her community collaborators, Chief Albert P. Naquin, the Traditional Chief of the Isle de Jean Charles Band of Biloxi-Chitimacha-Choctaw Indians from coastal Louisiana. Chief Albert, as you will see, is a seasoned veteran of encounters with well-meaning scholars, students, NGO staff, and government officials. His valuable perspective about what works and doesn't work in community collaboration—and what his ideal type of engagement looks like—is important for academic practitioners to hear, and he also reflects back on Peterson's work as being a valuable approach. I am very grateful that Chief Naquin took the time to collaborate with me on his piece.

When Richard Campanella and I discussed this book project early on, he said, "If there isn't anything about social media in the book, it will be almost immediately obsolete." I agreed and, obviously, took his words to heart in deciding to research that subject. I chose Twitter as the digital social platform to focus on because it is the one I am most familiar with, and I also focused on the biophysical sciences, that being my specific area of interest in civic engagement. What I found is that social media, and Twitter in particular, are being widely used by scientists, and this state of affairs is being problematized and discussed by scientists and social scientists alike. Not only is social media a powerful tool for scientific public engagement, but it is really transforming the way we understand civic engagement as a whole.

Chapter 5

Effective Engagement: Critical Factors of Success

Janice Cumberbatch

Introduction

Having spent many years advocating various forms of participation in natural resource management as well as various social development projects, I decided that I would make it the focus of my doctoral research. The challenge was deciding what the research question would be. That was resolved when my thesis advisor asked me "What do you *really* want to prove about participation?" to which I responded, "Whether it really works or not." That eventually became more refined into an investigation of what makes participation effective. It seemed to me to be a very important question since participation or civic engagement was advocated as providing many direct and indirect benefits to the public. Numerous researchers have argued that it strengthens democracy (Langton 1978; Renard 1991; Rudqvist and Woodford-Berger 1996; Borrini-Feyerabend 1997; Warner 1999). Others posit that participation empowers beneficiaries and improves the quality of project outputs by uncovering new understandings, securing buy-in for decisions, and limiting delays, mistakes, and lawsuits (Uphoff 1991; Renard 1991; Borrini-Feyerabend 1997; Burby 2003; Kessler 2004). Consequently, many projects have a participation component into which money, human resources, and time are invested. Therefore, I set out to conduct a detailed analysis of the concept of participation to better understand how it was being implemented so that I could make suggestions that would improve its effectiveness and make good on the investment of time and resources that citizens contribute; as well as the time and finances that various public and private sector agencies and financial institutions commit to these projects.

The doctoral thesis was ultimately focused on the following research question: *What are the factors that enhance people's participation and by so doing, contribute to the successful achievement of a project's goals and objectives?* The goal of the study was to generate a matrix of factors that must be present for participation to be effective, and arrange these factors into easy to use checklists that would assist in the design, monitoring and evaluation of participatory projects.

Methodology

Coming up with the research question was a relief until the difficulty of developing a methodology for monitoring and evaluating the effectiveness or success of participation became apparent. The nature of participation creates this difficulty because firstly, it is a dynamic process that must be evaluated over time; therefore, taking a onetime snapshot of participation in any given project would not have provided a realistic and true measure of its effectiveness. Secondly, participatory projects do not occur in a laboratory or a vacuum protected from social, political, economic, and environmental factors, and any of these can help or hinder the participatory efforts and had to be accounted for. In addition, there is the fact that participation is understood in vastly different ways in development projects and programs. It impacts not only on the project or program but also the different stakeholders in different ways; and it can be both a means to help achieve project efficiency, effectiveness, coverage and sustainability as well as an end to promote stakeholder capacity and empowerment and therefore, must be monitored and evaluated from these two different perspectives (Karl 2000).

In arguing this point Oakley stipulates that the evaluation of participation requires two simultaneous approaches; quantitative measurement based on numerical values leading to judgement, as well as qualitative description leading to interpretation (cited in Karl 2000). So for example, there could be a quantitative measure such as the number of people who participate in a project, coupled with a qualitative indicator such as the extent to which the people who participated demonstrated leadership throughout the varying stages of the project. Marsden and Oakley took this discussion further by recommending the need for the development of valid criteria for understanding the nature of participation in a project; as well as the need to establish a set of indicators to give form to these criteria (cited in Karl 2000, 14). However, most researchers opt to select evaluation factors on a case-by-case basis according to the specifics of the individual projects.

In the *Resource Book on Participation*, the Inter-American Development Bank (IDB 1997) offered a set of specific indicators for assessing "when participation may be relatively easy" and "when participation may be relatively difficult" and these were reviewed in my study. However, given the complexities involved in assessing participatory processes and the absence of an agreed set of criteria, I chose to focus my attention on the experience of practitioners to identify the qualitative factors that have repeatedly, in various settings, contributed to the effectiveness of participation. I obtained these experiences using three methods of data collection: a literature review, a focus group discussion, and a survey. I surveyed literature originating in the 1950s on community development, up to current day research findings and reports of participatory efforts, to determine trends in theory and practice.

The focus group session comprised six practitioners from the gender studies, adult education, human resource development, and natural resource management fields, each of whom have been involved in participation for more than two decades. The data collected from the literature, in addition to the focus group session, were used for the initial identification of the factors that make participation effective. I then further assessed the factors obtained from these sources during the questionnaire survey. The survey sample comprised 85 practitioners who had worked in a wide range of geographic regions across the globe: Africa, Asia, the Caribbean, Europe, North America, and South and Central America. More than half of the sample, 64.8 percent, had been using participatory methods for more than 10 years, and 30 percent had been using these methods for over 20 years. They had employed participatory approaches in a diversity of fields as reflected in Table 5.1. Some of the projects and activities included the organization of charcoal producers to manage mangrove resources St. Lucia; mobilizing a community in Trinidad and Tobago to protect sea turtles through the development of a community-based education and tour group; networking communities in Colombia to plan, implement and evaluate family welfare and health services; and in environmental impact assessments in the oil and gas and forestry sectors in Canada.

I used the questionnaire to identify the factors that made participation effective in two ways. First I asked the respondents' to confirm whether or not the factors that I had identified from the literature and the focus group session had also worked in their experience. This was achieved primarily through Likert scale statements (http://www.socialresearchmethods.net/kb/scallik.php). Secondly, I used open-ended questions to allow the respondents to identify additional factors that they had found to make participation effective.

Table 5.1   Survey respondents' areas of work

Areas of work	Percentage of respondents	Areas of work	Percentage of respondents
Academic	22.4	Gender	5.9
Arts and culture	9.4	Health	12.9
Agriculture	11.8	Law enforcement	5.9
Coastal zone management	23.5	Marine resources	27.1
Community development	36.5	Religion	2.4
Education	25.9	Sports	3.5
Environmental Management	51.8	Tourism	20

Note: N=85

Note: Some respondents' work fell into more than one category, thus the combined percentages listed here equal more than 100 percent.

The Seven Factors of Effective Participation

In the research I defined participation as "the process whereby individuals and groups are afforded opportunities through specifically designed activities to be actively involved in any or all of the stages of conceptualization, design, planning, implementation, monitoring and evaluation of projects, programs or policies." For the sake of simplicity, I use the word "projects" to represent projects, programs, and policies throughout this chapter. My research revealed seven factors that must be addressed for participation to be effective. These factors were:

1. The definition of participation
2. The potential participants
3. The implementer or facilitator
4. The funder
5. The type of project
6. The social context
7. The resources

The factors pertaining to the participants, number two in the list above, emerged as those most likely to cause the success or failure of a participatory project. However, in each category issues were identified that could significantly affect the outcomes of participatory process. I then organized the seven factors into checklists, designed to encourage practitioners to ask a series of questions to ensure that everything pertaining to each factor has been or can be accounted for within the project. Consequently, the checklists were intended to be used to design as well as to monitor and evaluate projects geared towards civic engagement.

Definition of Participation

One of the first steps that should be taken when planning or undertaking civic engagement is to determine what participation means in the context of the project. Kavanagh points out that "the term has become so popularized that it has lost any real precision of meaning" (Sadler 1980, 1). Likewise, Renard states that participation is a "buzz word" in the jargon of planning which has become increasingly popular, but despite the popularity, the concept remains relatively unclear, and the word is used to describe activities of diverse kinds (Renard 1986, 1).

The literature I reviewed revealed three central themes embedded in the varying definitions of participation, namely:

1. Purpose: participation used as a process to achieve a goal, or as a desired outcome within a project, in other words it can be both a means and an end in a project.

2. Techniques: there are numerous techniques to be employed to achieve participation.
3. Levels: there are varying levels of participation that can be employed at various stages in the project cycle.

In addition, the focus group participants and the survey respondents added a fourth factor to be considered in defining participation, the source—who is requesting the engagement (Table 5.2).

Both the participants and the implementers of a project need to be extremely clear about the purpose that participation will play and at what stages in a project. Clayton takes the position that participation is both a means and an end, stating that participation can take place in different phases of the project cycle and at different levels of society, and take many different forms, ranging from contribution of inputs to predetermined projects and programs, to information sharing, consultation, decision making, partnership and empowerment (Clayton et al. 1998). Most writers' work on the subject supports this dual role of participation. Oakley, for example, interprets popular participation along three broad lines, in which it is a process through which persons can contribute to, or be organized in a project, and it is also an end result in that it fosters the development of skills and self-reliance (Oakley 1991, 6). The Caribbean Natural Resources Institute (CANARI) also supported the dual role of participation, defining it as both a process and an outcome through which concerned stakeholders become actors in decision-making that affects their lives and their communities (CANARI 2005). Thus as Sadler says:

> ... public participation is many-sided. It can be used to organize people and to get their input—ideas, skills, and other resources—into a project; it can build capacity and enable people to empower themselves; it can be used to involve people in decision-making; and it can be used in the process of monitoring and evaluating. (1980, 1)

Table 5.2   Checklist for defining participation

Purpose	Is participation to be the process, the result or both?
Technique	What methods will be used to engage stakeholders or participants?
Level	What level of participation is required at each stage of the project?
Source	Who wants participation? The potential participants, the implementer/facilitator/ or the funder?

Given the wide diversity of possible interpretations of participation, agreement on its purpose in a project is therefore critical because of the possible divergence among the expectations of the various players engaged in the participation activity. Therefore, Kessler (2004) advises that the goals for a participation process should be established early and communicated clearly. Bryson and Quick (2013) concur with this perspective, stating that clarity about the purpose of the participation process can help avoid unnecessary or unwise expenditures of effort and resources, problems measuring the outcomes of the effort, or challenges to the legitimacy of a participation process because of conflicting ideas. For example, if the agency that is financing a project for the development of sanitation infrastructure views participation as a means to an end, the expectation is that at the end of the project timeframe, the toilet and water facilities in a given community would be installed by the residents of the community. Conversely, if the process of engaging the residents in the laying of the pipes and the building of the toilets is seen as an end in itself by the implementing individuals and groups, the latter would consider that success has been achieved if the participants' level of awareness of health issues was raised, even if the toilets were not completed in the timeframe allocated. In this example the financier would believe that the funds were inadequately invested and perhaps not provide funding to that set of individuals or agencies in the future.

In this regard it is noteworthy that the majority, 84.7 percent of the survey respondents, agreed that a participatory process can be successful even if the project goal is not achieved. One respondent even stated that success in participatory processes occurs: *"When there is recognition that not only the end results but the processes used to achieve those end results are equally valuable and instructive."* Therefore, clearly stating whether participation will be used fundamentally to drive the process or whether a few of the project activities will incorporate some participatory techniques can lead to shared and understood expectations. This also ensures that the appropriate techniques can be selected for use at the optimal stages of the project cycle.

Techniques give effect to participation; they are selected depending on what is to be accomplished. Connor (1980) for example recommends that the technique be determined according to the stage of the project in which it will be used. An alternate approach recommended by Modak and Biswas (1999) is to classify the techniques by the function they would perform or facilitate, i.e. media techniques, research techniques, political techniques, structured group techniques, and large group meetings. Whether they recommend selection of technique by project cycle or function, the writers agree that they must be selected according to the goals and objectives of the participatory process, as well as an understanding of the local situation, in other words, the target audience, and the information to be shared or obtained.

This is sound advice, especially in light of the wide variety of possible techniques that can be employed to engage participants. In my survey of 85 practitioners, a dizzying array of more than 300 examples was offered, in response to the question, "What methods do you use to engage stakeholders?" These examples ranged from easily recognized processes such as participatory action research and co-management, to a variety of gatherings including meetings, conferences, focus groups, and buzz groups, to activities that, while not automatically participatory processes or methods, could be undertaken in a participatory manner, for example surveys, information sharing, and research. Some respondents also listed specialized methodologies for facilitating participation including the Technology of Participation and the ZOPP Methodology by OECS/GTZ. This confirmed the earlier point that participation has become increasingly popular, but the concept remains relatively unclear, and the word is used to describe activities of diverse kinds (Renard 1986).

Thus, despite the fact that participation is advocated by funders, NGOs and CBOs, governments, and academic institutions, I found that it still had no common language. Therefore, every effort must be made to ensure that the techniques selected are understood by all the participants and can achieve the goals of the project, because as Kessler clearly states: "different participatory mechanisms lead to different levels of involvement, with some merely facilitating information sharing and others providing opportunities for real deliberation" (Kessler 2004, 19).

Another critical aspect in the definition of participation is that there are varying levels of participation. This was discussed in the focus group and was the subject of much of the literature. Historically, one of the more popular approaches utilized when describing and defining participation has been to establish a continuum. For example, Rudqvist and Wodford-Berger suggest that the various definitions of participation may be viewed along a continuum from the more far-reaching or profound with respect to empowerment, influence and control on the part of grassroots participants, "intended beneficiaries" or other key primary and secondary stakeholders, to more conventional conceptions where donors, agencies or project staff still essentially retain decision-making power and control with respect to key planning and functions (Rudqvist and Woodford-Berger 1996, 11).

One of the earliest writers to acknowledge this continuum of participation was Arnstein (1969). She concluded that participation had a number of levels and devised an eight-rung ladder of participation starting at the lowest level with manipulation and building at the highest to citizen control. Such a continuum or ladder is quite useful in identifying the level of participation to be employed in the project. However, the benefits of this approach have led over the years to the development of many variations of the ladder of participation, and there is therefore still no single classification that can be used as a basis for understanding. For example, Connor (1980) proposed a continuum that ranged from persuasion to self-determinism; the Department For International Development (DFID) also created a ladder of participation ranging from the lowest level of participation, i.e., co-option, coercion and consumption, through stages of compliance, consultation, and cooperation, up to collective action or co-learning (DFID 2004). CANARI (2011) has developed a continuum of participation in natural resource management that ranges from consulting to transferring authority or responsibility completely. More recently, practitioners turn to the International Association for Public

Participation's widely used spectrum of participation, with levels of participation that range from informing to empowering stakeholders (http://www.iap2.org/associations/4748/files/IAP2%20Spectrum_vertical.pdf).

What I found to be consistent across the various ladders and continua is that as the level of participation increases people are expected to invest more of their resources, and that the resources that people put into a project also appear to differ depending on the degree of participation. Thus at the lower levels, people are expected to merely acknowledge the project or program. As participation really starts, they are expected to increase their inputs of time and ideas, and at the highest levels they are required to manage and take responsibility for projects and programs.

A critical issue emerging from the many possible levels of participation is the acceptance that not all projects may require the highest level of participation, and that practitioners should declare the level of participation that is required to achieve their project goals and objectives. Kessler advises that managers evaluate what level of participation is appropriate to their situation, given their stated goals, and plan accordingly (Kessler 2004). By doing so, it is made clear to all concerned, especially those who are expected to participate and those who will be paying for the project, how much time, money and other resources will have to be invested, and more importantly, what level of return can be expected from the investment. In short, potential participants would better understand when they are only being asked for information, but should not expect to have an impact on the decision-making process.

The final factor in relation to defining participation is the source of the project. The question to be asked is whether the project and its participatory approach is being requested by the potential participants or is it being introduced by an external agency—for example, a government-conceived and designed project funded by public funds or those of an international lending agency? The focus group participants stated that this presented significant implications for the effectiveness of participation. In the case of the externally conceived project, it has to be sold to the potential participants because it may define a problem and a solution that they have no "felt need for" or are not "ready for." This being the case, the participation may be non-existent or lackluster, and appropriate mobilization techniques in the early stages of the project cycle, and adequate funding to support this mobilization for as long as necessary, may be required.

For example, it was felt that a participatory approach would benefit the redevelopment of the Folkestone Marine Park in Barbados which was being funded by government and implemented by a group of Canadian consultants. However, the local stakeholders had always been excluded from the decisions made about the marine park. Therefore, we decided to design a project that would involve the stakeholders and allow them to work towards a consensus on the changes to be made in the park. We started by sending letters to invite everyone within the project zone as well as other related stakeholders across the country to a meeting at which we shared information about the project and they began to share their concerns about the park. This was followed by meetings with individual stakeholder groups such as the fishers, divers, boat and cruise operators, hoteliers and residents. We took a boat trip with the divers and the fishers to understand and address some conflicts that were occurring at local dive sites; and finally a series of roundtable sessions were convened at which the various groups and individuals were able to express their concerns and negotiate revised park rules and boundaries.

When the project originates from within the community, the participants are more likely to be willing to get involved. However, willingness is not capability, and so it may still be necessary to incorporate appropriate techniques to ensure that capacity is built to allow participants to become effectively involved in decision making and implementation.

Factors Influencing Project Success

The Potential Participants

Potential participants or the intended project beneficiaries are those persons who are expected to invest time, effort and other resources to assist in the development of the project, and who, upon its implementation, will reap benefits. It could also be the group that may be negatively impacted by the project and therefore by virtue of this risk have a legitimate claim to be involved. This category deliberately excludes what I call the "implementers or facilitators;" in other words, the representatives of the agency that is facilitating the process, whether academic, public or private sector, non-governmental organization (NGO), or community-based organization (CBO).

As mentioned earlier, the factors pertaining to the participants have emerged as those most likely to cause the success or failure of a participatory project. The literature, focus group participants and the survey respondents collectively identified 13 factors specifically relevant to the potential participants that would enhance engagement in projects and increase the likelihood of achieving project goals and objectives (Table 5.3).

Table 5.3   Checklist for potential participants

Identification	Who will be affected by the project?
Organization	Are these potential participants organized?
Demographics	What are the characteristics of the participants that are relevant to the project?
Resources	What resources do the participants have that are needed by the project?

Table 5.3   *concluded*

Agreement	Do the potential participants agree to the process?
Previous experience	Do the potential participants have any previous experience with participation?
Awareness	Are they aware of the issues that are to be addressed?
Felt need	Have they expressed a felt need for the project or for a participatory process?
Readiness	Are there any signs that they are ready for such a process?
Willingness	Are they willing to participate?
Capacity building	Do they have the skills to participate? Will the project build these skills?
Conflict	Are there hidden agendas? Will anyone try to sabotage the project?
Ownership	Is there evidence that they have bought into or feel ownership for the project?

The involvement of the "right participants" was identified as necessary to influence the effectiveness of participation. There was unanimous agreement across the literature, focus group and survey that persons who would be directly affected by the project should be engaged. However Bryson and Quick (2013) stated that one of the challenges in participation is how to ensure that the appropriate range of interests is engaged in the process, especially those normally excluded from decision making by institutionalized inequities. They therefore recommend that stakeholder identification and analysis are critical tasks to undertake to ensure that marginalized groups are at least considered and may have a place at the table.

For 25 percent of the survey respondents, the stakeholder analysis would therefore have to identify participants who were influential, in other words those with the social and political maturity or skills and expertise to mobilize others to join into the process. The analysis would also have to locate those persons with control over key resources—natural resources, information, knowledge—that are central to the success of the project. I found it interesting that influence and access to resources for these respondents were not synonymous with high levels of education, social status or any particular gender. Indeed they did not score demographics such as academic qualifications, gender, literacy and level of formal education, level of income, and social status very highly as factors that would determine success of participatory efforts (Table 5.4). This underscores the importance again, of selecting techniques for informing and engaging participants that would overcome issues of literacy, education, social status and gender.

Table 5.4   Respondents' view on the characteristics of potential participants

Stakeholders' characteristics	Always critical (%)	Sometimes critical (%)	Never critical (%)	No response (%)
The stakeholders are illiterate	5.9	30.6	49.4	14.1
The level of formal education of the stakeholders	4.7	43.5	44.7	7.1
The stakeholders have high social status	0	35.3	56.5	8.2
The stakeholders have low social status	1.2	30.6	61.2	7.1
The stakeholders have a high level of income	1.2	29.4	60	9.4
The stakeholders have a low level of income and may even be considered financially impoverished	3.5	34.1	55.3	7.1

Note: N=85

There were some factors that were deemed of great importance where the participants were concerned. Thus for example, 72 percent of the survey sample deemed it always critical that there be agreement among the participants to the participatory approach. According to Bryson and Quck (2013) practitioners should establish with both internal and external stakeholders, the legitimacy of the process as a form of engagement and a source of trusted interaction among participants. In addition to agreeing to the process, 85.9 percent of the survey respondents said that when the participants have a felt need for the project, and readiness for participation (62.3 percent) this made a participatory approach work better. In the absence of need or interest, the focus group members suggest that people's participation can still be enhanced or encouraged by employing a well-timed external intervention, focused on mobilizing interest in, and a need for the project. However, they cautioned that the potential participants have to be open for this. Focus group members felt that it was easier to stimulate a group that has had positive previous experiences of participation, and therefore possesses a desire to participate and to experience change, as opposed to a group that embraces and protects the status quo. This was supported by the survey respondents, 74.1 percent of whom felt that groups which had negative previous experiences with participation were more difficult to engage. Also in support of this, Chen (2010) argues that when these experiences are positive, commitments increase and trust builds.

A well-organized group with effective leadership was also found to be better for encouraging participation by 20 percent of the survey respondents. Comments from these respondents indicated that conducting participatory projects can be difficult where there is a large heterogeneous group of individuals with diverse interests, who do not belong to any formal or informal group and lack leadership.

I also identified a number of external factors that can influence the engagement of participants. Among these was the issue of trust: survey respondents stated that people who trust the external agency and feel secure about participating were more likely to get involved in the project. The focus group members raised the issues of the pressing commitments that people have on their time and resources that may prevent them from being involved. Therefore, they urged that an understanding of these competing factors would be vital for encouraging the involvement of some people.

The focus group also suggested that people need to understand the issues surrounding the project so that they appreciate the importance and benefits of their participation. Kessler (2004) suggests that participants who see their contributions as making a difference and perceive the process as legitimate and fair are more likely to comply with the result. To this end, it may also be necessary to build the capacity of the group to bring them to a level where they can participate. In this regard CANARI (2011) recommends that capacity assessment and building are integral aspects of participatory planning, and would entail identifying the capacities required and developing and implementing an effective training strategy.

According to the focus group participants, participation can work in situations where conflict exists, but only at a level where it is not causing severe rifts within the group and therefore precluding the ability to dialogue and come to agreement. However, if the conflict is deep-seated and affects resources that may be critical to the implementation of the project, it may prove to be destabilizing to participation. Ultimately, if potential participants can be committed to the process of participation, feel ownership of the project, and can indeed be validated by the project, this augurs well for the effectiveness of participation and the successful involvement of the participants and the achievement of the project goals. Yet if participation comes at too high a cost to the participants in terms of their time, money or values, they may understandably opt not to be involved.

Implementer/Facilitator

The research revealed that the individuals and agencies responsible for initiating the process, i.e. the project facilitators/implementers, are also critical to the effective implementation of participatory projects. These persons or agencies may be public or private sector, NGO or CBO. They may also come from among the potential participants themselves, as in the situation where the project is designed by a local community. The participants may also be the funders. Factors that impact the effectiveness of participation from among the implementers are summarized in Table 5.5 below.

The literature and the focus group made it clear that the implementer/ facilitator plays a pivotal role in achieving effective participation within a group. These individuals are responsible for ensuring that there is proper planning, reporting and communicating for participation to take place and be successful within the project. The survey responses were very similar with respect to whether the implementer was government, NGO, CBO, or local or foreign consultant, suggesting that this was not considered as strongly influencing the effectiveness of participation; foreign consultants did have consistently lower scores however (Table 5.6). What was of greater importance was that these agencies/agents have previous experience with participation and in conducting similar projects. Teamwork and a willingness to share responsibility on the part of the implementers were also factors that encouraged positive results in these types of projects. The Inter-American Development Bank (IDB) indicated that difficult situations would be caused by implementers without experience with participation and limited capacity to implement participatory processes (IDB 1997). Kessler (2004) agrees with this and argues that when designing a participatory process, the capacity of both the lead agency and stakeholders should be assessed so that time, money, training and expertise do not become barriers.

Table 5.5   Checklist on the implementers

Identification	Who is the implementer? (government, private sector, NGO, CBO, consultant, participant group, other)
Agreement	Do they agree to or have a policy on participation?
Previous experience	Have they implemented similar projects?
Capacity	Is there a trained and experienced facilitator?
Control	Is the implementer prepared to share authority over the process and outcomes?
Planning	Is there a plan for implementing the participatory process? How will this be communicated to the participants?
Resources	Does the implementer have access to the necessary equipment, finances, expertise?

Table 5.6   Respondents' views on the characteristics of the implementers

Implementers' characteristics	Always critical (%)	Sometimes critical (%)	Never critical (%)	No response (%)
The implementer is government	7.1	56.5	28.2	8.2
The implementer is a local consultant	7.1	54.1	31.8	7.1
The implementer is a foreign consultant	5.9	44.7	38.9	10.6
The implementer is a local NGO from within the community	11.7	55.3	24.7	8.3
The implementer is a local NGO from outside the community	3.5	56.5	32.9	7.1
The implementer is a foreign NGO	4.7	43.5	42.3	9.4
The implementer is a local CBO from within the community	12.9	63.5	16.5	7.1
The implementer is a local CBO from outside the community	4.7	56.5	29.4	9.4
The implementer has prior experience in conducting participatory projects	42.3	45.9	5.9	5.9
The implementer has substantial experience in implementing similar projects	28.2	58.8	7.1	5.9
The implementer is willing to share responsibility at all levels	45.9	42.3	5.9	5.9
The implementer is willing to share responsibility but will retain the right of final authority	22.3	54.1	15.3	8.2

Note: N=85

The focus group participants suggested that the presence of policies conducive to the practice of participation within the implementing agency provides a support base for the work to be completed. In their opinion, such an agency is likely to provide training as well as financial and other technical support for staff to be able to effectively conduct participatory work. On the other hand, facilitators from agencies that do not support participation may themselves also not support it, or be working in a vacuum, relying on their own limited resources to conduct projects.

To aid in the effective implementation of participatory projects, CANARI has established a framework for reviewing institutional capacity for participation. It seeks to provide facilitators and lead organizations with insights into what stakeholders require in order to be effective partners, and whether an organziation is equipped for participatory management (CANARI 2005). The framework has the following elements: worldview, culture, structure, adaptive strategies, skills, material resources, and linkages.

According to CANARI, the worldview is a coherent frame of reference that the organization or group uses to interpret the environment in which it operates and defines its place within that environment. This should include a clear vision and mission, which provide a rationale for all other aspects of capacity. Based on this perspective, organizations will not have the capacity to design and implement participatory approaches if they do not have a vision that places community members, stakeholders, or participants at the center of development and addresses issues of equity and sustainability. Similarly, the framework states that culture is a way of doing things that enables the organization or group to achieve its objectives, and a belief that it can be effective and have an impact. CANARI states that organizations require a culture that embraces principles such as a willingness to work with other stakeholders towards shared objectives. These elements of the framework—worldview and culture—are consistent with points made in the literature about the need for democratic social and institutional contexts that support participation.

The third element, structure, is defined in the CANARI framework as a clear definition of roles, functions, lines of communication and mechanisms for accountability. Adaptive strategies are practices and policies that enable an organization to adapt and respond to changes in its operating environment. Skills encompass knowledge, abilities and competencies; material resources include technology, finance and equipment, and linkages signify an ability to develop and manage relationships with individuals, groups and organizations in pursuit of overall goals. According to CANARI, all of these elements are required at the institutional level for participation to be effective.

Funder

According to the OAS (1999), a significant factor in the success of a participatory process is the availability of financial resources to initiate, fortify and/or continue participatory practices. On many occasions there is no distinction between the implementer and funder. However, the separation is made here for the purpose of analysis. Based on the information collected from the literature review and the focus group, factors that enhance the ability of funders to support participatory efforts are listed in Table 5.7 below.

The IDB (1997) stated that it was important to provide adequate funding to make participation effective. The focus group noted that funding never seemed to address critical issues such as the time it took to mobilize and make a set of people ready to participate. In the survey, 87.1 percent of the respondents agreed with the following statement: "The agency or individual who finances a project can influence the success or failure of a participatory process." This clearly shows that the respondents support the findings of the literature review and the focus group regarding the important role that funders can play in influencing the outcomes of participatory processes.

The main requirement of the funders therefore, is to ensure that there is an adequate supply of financing available to sustain the project activities in each cycle. Persons in the focus group and the survey went further however, and noted that funders also need to be sensitive and recognize that some activities, for example mobilization of a group to make it "ready for participation," require more time than they currently allocate or allow for in their planning and disbursing of funds.

Sometimes funders pre-determine that a project must be participatory. In some instances this may not be appropriate, and therefore there is also a need for funders to open dialogue with implementers and potential participants in the design stages of projects to determine how extensive a role participation may play on a case-by-case basis.

Table 5.7   Checklist on the funders

Identification	Who is funding the project? (e.g. government, IFI, NGO, CBO, other)
Agreement	Does the funder agree to the participatory approach?
Resources	Will the funder provide the required funds to support the project? Will they agree to the timeframe?

Table 5.8   Checklist on the type of project

Goals and objectives	Is a participatory process necessary to achieve the goals and objectives of the project?
Rules	Is there a shared understanding and agreement to the rules of the process used in the project?
Benefits	What benefits will the project provide? (e.g. direct or indirect, the beneficiaries, the nature of the benefit)
Technology	Are the technological methods used in the project compatible with the capacity of the participants?

Type of Project

The type of project is also a critical part of determining whether participation will be effective. Factors that influence the role that project design will play in the success of a participatory project are included in Table 5.8 above.

Thomas (2012) suggests that participation should be sought when it is required or when it is the only or most efficacious way of gaining one or more of the following: needed information, political support, legitimacy, or citizenship development. In this regard the survey respondents displayed a measure of ambivalence with 44.7 percent agreeing that a participatory approach should be undertaken in every project, and 52.9 percent disagreeing. Instructive were the types of projects for which the respondents indicated that they would *not* utilize a participatory approach:

- Projects where participation would have no impact: projects with a clear, mandate and definition, where participation will not make a difference to the implementation plan. For example, a decision to change the computer system utilized by the customer service center within an agency may lead to delayed response times, as the staff learn the system. In such a case all that is required is a public service announcement to inform the public of the likely delays.
- Scientific and precise technical studies: answering narrow scientific and technical questions; or situations where the outcome needs to be controlled, reproducible, verifiable and exact; or when the results of the study or project are for academic use only. Situations which meet this description include experiments in crop breeding, DNA analysis etc.
- Projects where there is no room for negotiation: where there is no real chance that the audience's participation would be likely to influence a pre-determined situation/outcome; or where the decision is not open to debate, in which case this is clearly an information only exercise rather than true participation in the decision. For example, most customers would likely prefer to be included in the decisions made by commercial banking institutions to increase interest rates, although, sadly, it is not likely to occur.
- When urgent decisions are required: for example when decisions have to be made quickly; or if a decision needs to be made within a certain deadline. It is hardly likely that pilots will engage their passengers to discuss options when making split second decisions to avoid a plane crash or mid air collision.
- When there is a crisis: for example where urgent remedial action is needed; in an emergency situation; or in a crisis situation. First responders to hurricanes, landslides, plane or train wrecks are unlikely to engage in a participatory process at the time of the response. One would hope that such discussions would have taken place during a planning and preparedness phase, to ensure that all concerned, including community members, know what they are required to do during the crisis.

Once it is determined that a participatory approach will be utilized, then there should be a shared understanding and agreement about the project goals and objectives by all concerned (participants, implementers, funders); this extends to a set of rules that will govern the process of achieving these outputs. Rules about how the process will be managed and how decision making will take place provide a bridge between participation processes and organizational structures. For organizations, these rules are often embedded in legal mandates or commonly held beliefs regarding appropriate roles and responsibilities (Bryson and Quick 2013). In addition these rules must establish things such as: the role of participation in the projects; the roles and responsibilities of all involved; how benefits will be shared; and how reports will be made and disseminated among others.

Ninety-three percent of the survey respondents suggested that in order to involve the participants, the project may need to clearly show how direct or indirect benefits will accrue at the individual and collective level. Getting people to participate fully may also rely upon ensuring that the project design allows for the right level of participation at the right stages of the project and the use of appropriate techniques to get them involved. For example, the selection of appropriate techniques for stakeholder analysis in order to identify all the critical players; or formatting meetings to accommodate differences in culture, gender, or age are necessary steps.

If the project is initiated from within the group of potential participants there is a greater chance that it will enjoy full or substantial involvement. However, there is still the need to ensure that project inputs are compatible with the local context and that they are not alienating to the group. For example, a group of largely illiterate persons will not respond well to written notices and reports about the project.

The IDB (1997) stated that one of the factors that make participation easy is the type of technology used in the project. In the survey, technology was considered sometimes critical by 62.3 percent of the respondents. Wang and Bryer argue that participation processes that engage the public can be significantly enhanced by the use of information, communication, and other technologies (cited in Bryson and Quick 2013). These technologies include public participation geographic information systems, computer generated visualizations, interactive Web sites, keypad voting, and strategy mapping tools (Bryson and Quick 2013).

Social Context

The project occurs in a context—social, economic, political, cultural, geographic—that also has a bearing on the effectiveness of participation. Factors influencing effectiveness in this case are summarized in Table 5.9.

The focus group members suggested that having a culture of participation would have a positive impact on the effectiveness of participation. The literature review confirmed that participation is grounded in democracy and gives effect to it through the involvement of persons in the activities that affect their livelihoods.

For example, Davidson and Stein (1988), writing on participation in health care in Peru, concluded that among the obstacles to participation was the pre-existing internal hierarchy, in other words, the nature of the power structure that dictates relationships between social and hereditary groups can either facilitate or block participation. The IDB (1997) also found that difficulties with participation occurred in social contexts where there was no democracy; where the local authorities were unwilling to relinquish control and power; and where the level of bureaucracy was inflexible.

Table 5.9   Checklist on the social context

	Checklist: Social Context
Culture	Is there a culture of participation or a history of democracy?
Local authorities	Do they agree to the participatory approach? Will they share authority?
Bureaucracy	Is there any legislation, regulations or policy that addresses participation in this jurisdiction? Red tape? Effective communication mechanisms?
Change	What changes are the project likely to cause to the social/political/economic or cultural situation?
Conflict	Are there any existing conflicts among the local authorities? Will it help or hinder? Sabotage?
Geography	Are the potential participants geographically isolated?

It is not surprising therefore that 75.3 percent of the respondents agreed with the statement: "Participation works better in locations with a culture or history of participation" thereby confirming that the presence of democratic institutions and practices would enhance the effectiveness of participation. This is so because an environment of democracy facilitates a number of the other factors that are required to make participation effective, such as legal and regulatory frameworks and policies that support participation, adequate financing of the local governance structure, as well as forums to encourage the dialogue necessary for communication to take place among the various stakeholders. In short these factors signify that there is a culture supportive of participation, free of the level of political conflict and interference that would undermine participatory efforts.

Seventy-three percent of the respondents thought that it was always critical for the local authorities to agree to the participatory approach, confirming that the buy-in of the local authority is particularly important. Government or the ruling authority is seen as having the capability of either positively or negatively impacting the effectiveness of participation. To be positive, ruling entities must support participation and be willing to relinquish some of their power and share responsibility with other stakeholders, even though they will retain final authority as the governing body. For this to happen, normally severe bureaucratic systems must have in-built flexibility, and must be open to external approaches in order to articulate with the less stringent systems at work in CBOs and NGOs with which they share responsibilities in a participatory context. The ruling authorities must also have a history of keeping their promises, since failure to do so breeds mistrust which is counterproductive to a participatory partnership.

The research has also suggested that being sensitive to cultural differences within a group is the key to successful participation. Anecdotes from researchers who recall situations where facilitators convened meetings at times inconvenient to their potential participants attest to this. For example, setting up meetings on Friday evenings when a large proportion of the potential participants are members of the Seventh Day Adventist Church whose Sabbath commences on Friday evening, or expecting females to speak freely in an open forum with males present in a culture where it is not acceptable for females to speak in such settings. Geographic isolation is also considered to be a factor. In this case it is critical to ensure that potential participants are not excluded because of the difficulty that might be experienced in accessing them or making the process accessible to them.

Resources

Projects will not succeed if the required resources are not available. Based on the information gathered during the study, the resources required to ensure that participatory projects are successful are summarized in Table 5.10.

Table 5.10   Checklist on the resources

	Resources
Time	Has enough time been allocated for the implementation of a participatory process?
Finances	Is there enough money to implement a participatory process?
Human	Are the required human resources available?
Technology	Is the required equipment and technology available and accessible?
Facilities	Are they available and accessible?

Time emerged as the most important resource. This was the belief of the focus group supported by statements in the literature. In the survey 82.3 percent of the respondents also stated that time was always critical. For 81.2 percent and 72.9 percent of the survey respondents respectively, human resources and information were also always critical. Interestingly only 51.8 percent of the survey respondents thought money was always critical. Technology was seen as somewhat less important, but was considered sometimes critical for 62.3 percent of the survey respondents.

When we asked, "What factors help in the determination of when to use participation?" 22 of the survey respondents discussed resources, out of which eight specifically identified time. The focus group generally felt that time is greatly needed because participatory methods take more time to implement than non-participatory projects. In addition, five survey respondents identified human resources, specifically the specialized expertise and experience needed to enrich the planning and implementation process, and three responses indicated that financing was among the factors that they found to be necessary when deciding to use a participatory approach.

In considering when not to implement a participatory process, the survey respondents once again identified lack of time as the biggest constraint. They specifically stated that short time frames, tight timelines, and situations with serious time constraints made participation more challenging. Other responses dealing with resources also pertained to limited human resources.

Consistent with these results, the absence of time was the subject of 16 of 31 survey responses that dealt with resources in response to the question of what factors made the implementation of participatory processes difficult. Another nine of the survey responses identified the absence of funds, and three mentioned a lack of information.

That time emerged as all important is not surprising, since participation is time-intensive, requiring facilitators and implementers to work at times that do not fit within the normally accepted 8:00 a.m. to 4:00 p.m. timeframe. Rather, participatory efforts on the part of implementers spill into after-hours, late into the night for meetings, and on weekends and holidays. Significant investment of time is required of participants, who must attend meetings and other project-related events.

Participation is human-energy-intensive. Despite the diversity of definitions given when asked how it is done, there is one consistent factor; it is done by people with people. It requires both the facilitators and the participants to share information and skills and will not be effective if adequate locations are not available for project activities such as meetings and workshops.

Critical values

Finally, I found that there was a set of underlying values that the practitioners felt impacted the effectiveness of participation. These are listed in Table 5.11

The survey respondents and the focus group participants stated that it was important to be genuine, open, honest, consistent, fair, and organized when working with communities or stakeholders; and they felt that it was important for implementers to admit when they did not have all the answers. These practitioners believed that these attributes would help to establish the much-needed sense of trust and cooperation necessary for participation to work. Openness, in other words the willingness to disclose facts, transparency and fairness of all parties was also required. Fundamentally, trust and respect for all involved, especially for the leader/facilitator was considered very important.

These values affect all the players involved—the potential participants, the implementers and the funders. They embody the essence of what participation is all about: players who cooperate with each other and are committed to achieve the task at hand; the importance of trust, honesty and transparency among those who have to work together in the participatory effort. Effective participation is therefore fostered in democratic settings where players respect each other and when there is openness, equity and fairness in their dealings with each other.

Table 5.11   Value underlying successful participatory efforts

Underlying values		
Accuracy	Being genuine	Completeness
Cooperation	Comprehensive	Confidence
Consistent	Compassion	Commitment
Democracy	Empowerment	Equity
Goodwill	Kindness	Transparency
Honesty	Horizontality	Fairness
Predictability	Openness	Trust
Power delegation	Respect	Organized

Conclusion

The literature, focus group, and respondents present a position that participatory approaches can be used in any type of project, even though it may sometimes be difficult to implement. This reinforces the importance of having a mechanism for isolating the critical factors of success, so that these can be emphasized to enhance the likelihood of project implementation and achievement of goals. To this end, I placed the factors identified during the study in a checklist, which can be used to enhance project design, monitoring and evaluation. Those charged with the responsibility for project design and implementation can use it to analyze all aspects of their project ahead of the commencement of the project, thereby identifying project components that could weaken their ability to implement the participatory approach and so allocate their resources to strengthen these areas.

Funders can use the checklist to understand the nature of the projects that they are supporting and to ensure that the funds that they allocate, and the expectations that they have for the return on their investment, are compatible. The checklist offers funders a way of analyzing the potential participants in a project, as well as the proposed approaches to the project so that they can better understand why more time and finances might be required to achieve the required outcome. Alternately, they can also determine when their resources would not yield acceptable benefits. The checklist makes it clear what resources are required from the potential participants versus what resources they have available. In this way there should be a reduction in unrealistic demands on the participants. An added advantage of the checklist is its flexibility. It can be used during and after project implementation, so it is good for both monitoring and evaluation.

The unique nature of the checklist is that its focus is on participation. This is important because as the literature and the focus group participants confirmed, most processes of project evaluation focus on the activities and the outputs, or the use of the resources, but seldom on the actual effectiveness of participation. However, since resources are being expended to implement projects in a participatory manner, it is critical to assess whether participation is working, and this checklist offers the means to do this before, during and after a project. Moreover, since the checklist focuses on participation in its generic form, and not on its application to any specific type of project (for example, resource management, education, planning, or policy) it is of universal usage in project design and management.

In summary therefore, there are numerous factors that influence the effectiveness of participation. These factors pertain to the way participation is defined in the project, the potential participants, the implementers, the funders, the project design, the social context in which the project is taking place, and the resources available to support and implement the project. While a participatory approach can be used in any type of project, there are conditions under which it is very difficult to implement. It is under these conditions that the factors which influence effectiveness need to be fostered to increase the likelihood of project success. I arranged the factors in an easy to use checklist that I hope will assist in the design, monitoring and evaluation of participatory projects.

References

Arnstein, Sherry R. 1969. A ladder of citizen participation. *AIP Journal*, 34: 216–24.

Borrini-Feyerabend, Gloria, ed. 1997. *Beyond fences: Seeking social sustainability in conservation.* Gland: World Conservation Union (IUCN).

Bryson John M. and Kathryn S. Quick. 2013. Designing public participation processes: theory to practice. *Public Administration Review* January–February 2013

Burby, Raymond J. 2003. Making plans that matter: Citizen involvement and government action. Special issue, *Journal of the American Planning Association*, 69: 33–49.

CANARI. 2011. *Facilitating participatory natural resource management: A toolkit for Caribbean managers.* Laventille: CANARI.

CANARI. 2005. Governance and Civil Society Participation in Sustainable Development in the Caribbean. *CANARI Policy Brief,* No.7 August 2005.

Chen, Bin. 2010. Antecedents or processes? Determinants of perceived effectiveness of inter-organizational collaborations for public service delivery. *International Public Management Journal,* 13(4): 381–407.

Clayton, Andrew, Peter Oakley and Brian Pratt. 1998. *Empowering people: A guide to participation.* New York: The United Nations Development Program.

Connor, Desmond. 1980. Models and techniques of citizen participation. In *Public participation in environmental decision making: Strategies for change,* ed. Barry Sadler, Alberta: Environmental Council of Alberta.

DFID. 2004. *Tools for development.* http://62.189.42.51/DFID/FOI/tools/chapter_07.htm.

Inter-American Development Bank (IDB) n.d. *Resource book on participation.* Washington DC: IADB. International Association of Public Participation, http://www.iap2.org/associations/4748/files/IAP2%20Spectrum_vertical.pdf.

Karl, Marilee. 2000. *Monitoring and evaluating stakeholder participation in agriculture and rural development projects: A literature review.* Sustainable Development Department (SD), Food and Agriculture Organization of the United Nations (FAO) http://www.fao.org/sd/PPdirect/PPre0074.htm.

Kessler, Brianne Leigh. 2004. *Stakeholder participation: a synthesis of current literature.* Prepared by the National Marine Protected Areas Center in cooperation with the National Oceanic and Atmospheric Administration Coastal Services Center, September 2004. www.mpa.gov.

Langton, Stuart. 1978. *Citizen participation in America.* London: Heath and Company.

Modak, Prasad and Asit K. Biswas. 1999. *Conducting Environmental Impact Assessment for Developing Countries.* New York: United Nations University.

Oakley, Peter. 1991. *Projects with people: The practice of participation in rural development.* Food and Agriculture Organization of the United Nations (FAO), United Nations Development Fund for Women, World Health Organization, World Employment Program. Geneva: International Labour Organisation.

Renard, Yves. 1991. Institutional challenges for community-based management in the Caribbean. *Nature and Resources,* 27(4): 4–9.

Renard, Yves. 1986. Public Participation in the Management of Natural Resources. *Bulletin of Coastal Area Management and Planning CAMP Network.* October 1986: 1–3.

Rudqvist, Anders and Prudence Woodford-Berger. 1996. Evaluation and participation: Some lessons. *SIDA Studies in evaluation* 96(1). Stockholm: SIDA.

Sadler, Barry, ed. 1980. *Public participation in environmental decision making: Strategies for change.* Alberta: Environmental Council of Alberta and the Banff School of Fine Arts.

Uphoff, Norman. 1992. Fitting projects to people. In *Putting people first: sociological variables in rural development*, ed. Michael Cernea 359–95. New York: Oxford University Press.

Warner, Michael. 1999. Which way now? Choices for mainstreaming 'public involvement.' *Economic infrastructure projects in developing countries, Development Policy Review*, 17: 115–39.

Chapter 6

Participatory Action-Citizen Engagement: When Knowledge Comes from the Academy and the Community

Kristina J. Peterson

Given the gravity of the social and environmental issues facing high-risk coastal communities, it is essential that academics, agencies, and communities find common ground to respectfully listen, engage, and respond. For too many years, rural, out-of-the-way communities have been either left out of the process of studying problems with academics and agencies, or have been informed through such ineffective means as study results being deposited in pamphlet form at public libraries.

The scientific community and the government agency community are finding that the public's confidence in their work and leadership is diminishing, evident in factors such as fewer people entering the field of science, and the decline in public/government financial support for science (Raj Pander, unpublished presentation, Conference on Public Participation in Scientific Research, August 2012, Portland, Oregon). Likewise, there is often contempt by academics and agencies for local participation in research, viewing locals as incapable of doing or understanding science. Prospects seem remote for collaborative work to address future risks, such as climate issues, unless we find ways to bridge the gap between agencies, academics and citizens.[1] Doing so provides the opportunity for true public discourse and problem solving.

One path towards developing a common platform for discourse and trust building is through Participatory Action Research (PAR), or participatory citizen engagement. Through the use of participatory citizen engagement for issue problem solving, the citizen becomes a contributor to the development of knowledge with others and has ownership of the outcomes. Working in a mutual partnership with academics, citizens are respected for what they know about their issues of place or concern, and, as a result, the community participants' trust grows in the academic researcher, institution, or agency.

In this chapter, I will examine the use of PAR and how it has helped build bridges between academics, local communities, and government agencies by developing a public sphere for creating knowledge. Working through The Center for Hazards

---

1   Citizen is not being used as a political term but as a reference to a community member.

Assessment, Response, and Technology at the University of New Orleans and now the Lowlander Center, I have been engaged with several multidisciplinary projects embodying the philosophies of citizen engagement and participation in research, which I will describe here. Our approach has included aspects of traditional ecological knowledge, disaster recovery, and community resilience. I do not intend to focus here on the details of the individual projects themselves but on those elements that have contributed to trust building, mutual language, and knowledge exchange. My hope is that these examples will demonstrate methods that have helped build trust and confidence between citizens, political leaders, and academic researchers. I will share my reflections on our methods, and their implications for other participatory programs, as a possible strategy for working towards sustainable, resilient communities in a time of increased vulnerabilities and risk.

The Geographical and Cultural Context

Coastal Louisiana is an abundant storehouse of potential research projects: the region is home to multiple interwoven social and environmental challenges, including both naturally occurring and anthropogenic disasters and impacts. Communities are experiencing the fastest disappearing delta region in the world, mostly due to relative sea level rise and coastal erosion (List et al. 1997). In the time it takes to read this chapter, the region will have lost another football field of land. Some communities have already disappeared and some are trying diligently to relocate as the soil around them becomes submerged. Businesses and industry are relocating further north off the coast or moving to different states or countries. The seafood industry, and subsistence communities who gain their livelihood from Louisiana's marine and freshwater resources, are experiencing changes in the oyster, shrimp, crab and fish populations. Whether these issues are caused by multiple impacts such as changes in the estuaries from land loss, salt water intrusion, or oil spills and gas emissions, are not of primary concern to fishers: they are just wanting to be involved in environmental restoration and efforts to solve the problems they see as destroying their communities (Freudenburg et al. 2009; Rotkin-Ellman, Wong, and Solomon 2012).

Currently, NOAA's Office of Coastal Survey is reconfiguring new coastal maps of Louisiana. In a February 2014 interview with USA Today's Rick Jervis, Meredith Westington, Chief Geographer, indicated that the number of locations and place sites being eliminated from former maps—due to coastal land loss—is unprecedented. "This is the first I've seen it," Westington said. "I don't know that anyone has seen these kinds of mass changes before." This is the only place in the United States where Westington has to keep a running total on her desk of lost places. "It's a little disturbing," Westington said. "It's sad to see so many names go."

Amongst the places that are lost and ones that are "hanging on" are both historied and indigenous communities, some of which have existed in the region for several hundred years or more, with residents who are intimately connected to the land, water and living ecosystem. Given the harshness of the physical conditions, and the challenges to prospering in thick palmetto and cypress swamps and in hot and humid weather, coastal Louisiana bayou communities have learned how to live with—and in the presence of—risks and hazards. Many years in a dynamic relationship with the natural elements has given the communities intimate knowledge of their surrounding environment. The knowledge of plants used for medicinal purposes and as nutritional sources kept communities healthy without need of nonindigenous medicines or foods. The historical Louisiana coast was populated with people who had been sold off, pushed out, or who had escaped indentured servitude. In these communities—including people of African, French, Acadian, and Native American descent, amongst others—through such a close relationship with the land and place, rich traditions and robust cultures emerged, with deep histories that now tie them to their current disappearing locales (Brasseaux 2005).

The conundrum is that biophysical scientists will often propose fixing the loss of land through methods or projects that do not consider the residents, and often will not spend time with the communities to better understand local dynamics. Social scientists as well have studied the people, the culture, their values and lifeways without the collaboration or participation of the communities. For both biophysical and social scientists, a richer study and application of problem solving could take place if they engaged the wisdom and strength of the local communities, who have a knowledge base that can help address the dire conditions brought on by land loss and sea level rise.

The Tools We Have Developed

> When you first came to the community and asked to have a conversation, we didn't know what to say. University people usually have questions to ask us and get upset if we offer anything else. Just talking about things is different, it is sharing from the heart. (Community member, Jean Lafitte, Louisiana, interview by author, 2006)

NGOs, government workers, and academic researchers have surveyed many of the at-risk communities along the Louisiana coast numerous times over the last 10 years. The surveys are often multiple choice, and leave little or no place for the community to raise the questions or comments they might have on the subject under review. "When we get a survey now, we get together with friends for a Coke and decide how we are going to answer this one. *(laughs)* They don't seem to really care about us, so why not just make a game of it!" (Community member, Lower Jefferson-area, Louisiana, interview by author, 2009).

"It's time to move on from studies and make a difference. If we don't, there will be nothing left to study" (Community member, Lower Plaquemines, Louisiana, interview by author, 2011). Research that is initiated by the community and includes agencies and academics is centered on problem solving (Smith 1999). Research initiated by agencies, NGOs and academics might engage the community but is not always focused on solving the problems that are top priorities for the community involved. I have often received reviews of federal grant proposals I have submitted in which the reviewers criticize the proposed work for emphasizing action over research. Until there is attention paid to the ways research can help in solving real-world problems in a real-place laboratory within the midst of a community, we will not have truly transferable models nor will we develop workable, sustainable solutions. "How much money that has been spent trying to figure out why the coast is disappearing would have more than paid for the pipeline sediment diversions? [Sediment diversions channel sediments from one area to another and are designed to help build land in areas where land is currently being lost]. So instead of us having to go to another meeting to comment on another study, we could have been planting gardens of newly made land" (Community member, Lower Plaquemines, Louisiana, interview by author, 2013).

Over the past 10 years the researcher-community teams I have worked with have developed tools to assure the respect of—and true engagement with—communities, tools that include consideration of reciprocity of time, expertise and outcomes. Tools are merely resources and cannot be understood as the sole answer to appropriate and meaningful engagement. The tools I discuss here are meant only as examples or possible models for other researcher-community collaborations, not as recipes to be painstakingly followed without consideration of the unique characteristics of each setting and community.

Time is precious to historied subsistence communities like those in coastal Louisiana, who have a rhythm of time usage that fits their livelihoods and their family-community relationships. When negotiating a research project, the schedule of the *outsider team* needs to be aware that *their* schedule must fit into and match that of the community, *not the other way around*. The consideration of time includes accessible meeting places and having materials and research references available. In cases of "drop by" visitors such as reporters, the community can benefit from having briefing papers ahead of time as well as "bundling" reporters, college students, foundations, and so on, to coincide or overlap in their visits. Often, having the multiple reporters or groups present at one time affords better time and resource management for the community, and also offers a richer, deeper dialogue. Often when working with communities, I also find it important to bundle a visit with preexisting engagements, people, and other visitors so as to make full use of the time the community is giving. The community's life pattern must be respected, since some of their sustenance is based on lunar cycles, seasonal regulations, and exchange within the community.

Thus, when developing a working research relationship, compensation for time must be factored in: who will be compensated for what work, and when? It is often difficult to have funds set aside within an academic financial system to pay for local participants' time and on-the-ground community expenses.

Often researchers, especially after disasters, place an extra burden on the community simply by their presence. After one particular disaster, a graduate student I am familiar with wrote a successful federal grant proposal to cover the expenses of toilet paper, soap, and beverages while the university researchers visited with the community in a rebuilding project. In this circumstance, the community was extremely gracious with hospitality for the university team, but had limited resources to provide food, thus gift cards to a local grocer were given as a way to thank the community for their participation in the project, along with other compensation that had been agreed upon by the community members and the university team. The community appreciated the sensitivity of the graduate student to the burden that was being placed on the community by outsiders, and this helped build trust and a willingness on the part of community members to share their experiences. Often funds for these "extras" are not covered in grants (especially federally funded grants). Yet, there should be equitable distribution of resources in all phases of the research from start to finish, from inception to dissemination, in publishing and presentations. Continually over the past 12 years our teams have had to procure small foundation or faith community funds to enable local residents to fully participate in dissemination of findings and presentations at academic and professional meetings. Indeed, inclusion of the local community members in the dissemination and presentation of results has had deep impacts on the academic research partners, and has helped build bridges between communities, and academic and government institutions, as I will discuss later.

Reciprocity and mutual aid is a strong component of Louisiana coastal communities' resilience: people helping people. In a true community partnership, the outside research team must consider how they are going to reciprocate beyond the terms of the project. In one instance, a researcher brought some friends together for a weekend work-project to build a bus stop shelter for the community partner's schoolchildren A federal agency working on another project offered its expertise on brochures, and developed a brochure in conjunction with the community as a way to reciprocate. The more we can reach into our tool boxes of resources, and give in return to the community team, the stronger the bonds of understanding, appreciation, and collaboration will be.

The Private, Semi-Public, and Public Spheres

> From the 1950s through the 1960s and the 1970s, in the prevailing orthodoxies of development, professionals had the answers. In general, 'we' were right and 'we' were the solution. Local people, especially the poor were the problem, and much of the problem was to be solved by education and the transfer of technology. (Blackburn 1998)

One ongoing criticism of our technological world is the way life has become compartmentalized, including academic and professional life. With the increased specialization of specializations, each with its own technical language and protocols, crossing borders becomes more and more difficult, and the public sphere grows smaller and smaller. Full engagement of the public by government or the academy is oftentimes very limited (Fischer 2000; Forester 1989). As the public continues to be excluded from what should be a public sphere—of government or academic institutions—discourse becomes a closed private entity, a private sphere. Attempts by agencies and the academy to include the public have had mixed results. Often either people with power in the private sphere are assumed to represent the public, or there is a low level involvement of community members at public hearings (Cook and Kothari 2001). Town hall meetings, such as those historically held in the northeastern United States, have become antiquated as, in many regions of the country, more emphasis has been put on technical expert planning (Kunde 2007). Instrumental and technical knowledge in a closed system or private sphere run the risk of being representational knowledge void from the knowledge or ownership of all stakeholders (Gaventa 1993; Habermas 1991). Many people spend much of their professional lives in a variety of private spheres. Professions require the right credentials and adherence to the correct discourse and game-rules. Many academic meetings are conducted in the private sphere and require the proper credentials to attend. Some require invitations, registration fees, screened memberships, and the proper introduction or connections. As the private sphere is embraced, so are its prejudices, arrogance, and hubris.

As facilitators in the PAR process, we built a core team to work with—and alongside—communities of place and communities of interest, helping participants cross the borders and boundaries into the private sphere of professional academics and policy makers. This border crossing assistance included the professional and academic credentials, our network connections, our access to funding, and our personal histories of attending conferences and workshops in the private sphere. In several instances during our work in southeastern Louisiana, community members have engaged in spheres of influence usually reserved for academics and professionals such as academic meetings and professional associations. The communities have also allowed the outside academics to cross over into their lifeworld and lifework via invitations to venerated family and community events, and to participate in community advocacy.

Breaking Walls of Prejudice

Community participants interacting with academic and professional participants create, in Park's (1991) terms, relational knowledge sharing and building. It is in the building of relational knowledge that the private sphere can be expanded for the inclusion of others through the crossing of borders and boundaries. This expansion has not become what Habermas (1991) would term the public sphere because it is remaining only partially open and therefore not truly public. I call this transition the "semi-public sphere" since it no longer exists in the realm of either the private or the public spheres.

An example of this type of border crossing occurred when an academic researcher from North Carolina asked me and several of my colleagues in the New Orleans area if we had ever met members of a particular coastal community that, in fact, we did work with. The researcher was impressed by the community members' knowledge and expertise regarding coastal issues. She had met these community residents at a national meeting and wanted to inform us about their expertise, as a resource for our work. This indicated to me that the people of the community were being recognized for their own expertise, rather as a connection to me or as an extension of my academic institution. The community residents were part of the discourse of disaster problem solving in their own right, having both the tools of language and the desire to do problem solving. Unfortunately, however, many professional and academic meetings have barriers of membership, validation, degrees, language, or gender, as well as other restrictions that keep local experts and community members with wisdom apart from the discourse. These restrictions can lessen the value of the discourse: those who have gained entry are limited to dialogue with others like them, and to their own patterns of thinking and perception.

Silos are often created in the ways academics and professionals are trained in their disciplines. When academics are not trained in cross-disciplinary dialogue, it is difficult for some to comprehend or approach complex situations such as those faced by coastal communities of southeast Louisiana (Peterson 2010). This is compounded when academics and professionals come from a different socio-economic background, region, or culture than the community where they seek to work. A sole discipline or framework for action will not fix the complex suite of issues and problems caused by industrial growth and land loss in Louisiana.

Let us consider the example of communities that rely heavily on commercial and subsistence fishing. It isn't just a matter of having better marketing strategies for the fishers in order for communities to make ends meet. Their livelihoods and lifeworld depend on addressing multiple issues including—but not limited to—relative sea level rise, coastal erosion, deterioration of marsh quality harmed by oil and natural gas, stronger storms, lack of import taxes on seafood, and so on. When researchers and agency leaders—who do not live day-to-day with such layered complexity, or who mostly work within their own disciplines—visit coastal Louisiana communities, they are oftentimes so overwhelmed by the complexity that they deem the situation, people, or problem doomed. One university lectured stated, "Don't know what could help them, this is hopeless." However, researchers and agency leaders who are from locations—often outside the USA—that face similar challenges, and/or who work in multi-disciplinary venues, understand the context and have insight into the layered issues and the types of remediation that may be possible. Not finding the community hopeless, they also become creative, and attentive to the knowledge, insights, and experience of the local community members. For researchers and agency staff who see the problem as too big or complex, participatory work lends itself to working outside of the silos that are artificially created.

It helps the outsider to cross boundaries of discipline and their own prejudice so that they expand their thinking, creativity, and imagination.

Workers such as Chambers, Fischer, Forrester, and others are proponents of methods that will help form the necessary dynamics for participation of communities in the public sphere (Fischer 2000; Blackburn 1998; Forrester 1989). Participatory action strives for critical reflective knowledge through public discourse. Freire, Fals Borda, Park and others have often referred to this optimum state as "radical democracy" (Fals Borda 1991; Park 1993; Freire 1973).

Participatory Action Research as Radical Democracy

PAR is a democratically based research process that brings stakeholders together for problem solving that honors, centers, and reflects the experiences, knowledge, values, and hopes of the people most directly affected by issues in their communities. PAR is not so much a set of procedures as it is a philosophy and approach to gathering and using information. It is also a way to build and strengthen communities and their understanding of themselves, each other, and their relationships. It can be a powerful outreach, base building, and organizing tool that brings diverse people together to build understanding, which can create the environment for change. PAR is based upon the underlying concepts that participatory research, action, and reflection are emancipatory, democratic, and dynamic processes in which all participants must possess the values, humility, and courage to engage in collaboration. This form of cooperation aims to balance the power differential between people and promote the self-actualization of collaborating parties, and also to provide a forum for examining the interrelationships between partners and the effectiveness of the research process.

The primary purpose of PAR is to produce practical knowledge that is useful to people in the everyday conduct of their lives. A wider purpose of PAR is to contribute to the increased wellbeing—economic, political, psychological, spiritual—of humans and communities, and a more equitable and sustainable relationship with the wider ecology of the planet of which we are an intrinsic part. So, PAR is about working towards practical outcomes through knowledge sharing and knowledge building. Healthy action research emerges over time in an evolutionary and developmental process, as individuals develop skills of inquiry and as communities of inquiry develop within communities' PAR of practice. Action research is emancipatory; it leads not just to new practical knowledge, but to new abilities to create knowledge. Thus it is also a process of consciousness-raising, or conscientization, and is thus an educative imperative. Action research is, at its best, a process that explicitly aims to educate those involved to develop their capacity for inquiry both individually and collectively.

Presentation of Place

> A community partner exclaimed "I didn't know that you both were doctors!" while we were on an outing. It was wonderful that we were seen and treated as equals and not set apart as academics. (Academic partner, interview by author, 2012)

Throughout the literature on PAR not much has been written about the impact it has on the outside collaborator. But it follows logically that, in building a relationship of trust and expecting border crossings to occur, the outside collaborator will change and grow in some way (Peterson 2010). I have worked with one bayou community to investigate the impact PAR had on outsiders in a collaborative project. Our work has indicated that the outside collaborators did feel change that could be described through the use of my metaphor of border crossings. We juxtaposed the changes in the outside collaborators to Parks's knowledge typologies, to understand the type of change happening in the three knowledge types. The use of hermeneutics, while helpful in understanding the borders crossed and their process of conscientization in creating the public sphere, still does not give us adequate information on the variables that influence the process. We concluded that the following variables influenced the learning process of, or change in, the outside collaborators: presentation of place, validation of residents, duration of exposure, and the context of exposure. Here I will focus on presentation of place.

The place of encounter makes a difference in the outside collaborator's perception of the other, but place alone is not the only influence. The length of time and duration of exposure are also significant. The way people are validated, both from outside the community and by the community members themselves, influences border crossing and will create a change in perceptions. The background of the outsider is also a factor in the process of border crossing. The experience of the outside collaborator—whether the outside collaborator comes from a community that is similar or dissimilar to that of the people with whom she is interacting—will influence how the other is perceived.

The variable of place centers on how the community was presented to and by the outsiders. Some outsiders came to know a coastal location through student projects. For others, it was through fact-finding tours or trips responding to the devastation caused by the BP oil disaster, or Hurricane Katrina and other subsequent storms. Others were introduced to the community as a place where they could learn and understand the systemic issues of coastal land loss. The framing of the community, when presented to the outside, contributed to determining the manner of border crossing.

One example of the presentation of place can be found in a group of college students that came to the community as part of a spring break program. Even with advanced preparation about the community (using representational knowledge), and discussion with a professor regarding participatory work, the students seemed predisposed to understanding the community as one that was not whole, that was needy and required fixing. Many university work service programs have at their core a desire for the students to do something good for someone else during a break from their studies. Choosing an area of the country—southeastern Louisiana—perceived as "broken" as a result of disasters overpowered the advance work that was done on campus to foster a co-equal process of learning and teaching. When interviewed by our collaborative researcher-community team, the students talked about their desire to fix the problems of the community when they first arrived, but by the end of the week, with persistence, teaching, and interaction with community members, the students came to understand and appreciate the competency of the community. At the outset, the college students gave little credence to the ability of the residents to know or understand their own situation. After several days, the students began to understand the complexity of the issues and their innocence in thinking they were going to make everything better during one spring break week. The students' growing awareness of the community's knowledge was evidenced when they returned to their campus and raised funds to bring community members to the university to speak as experts during a weeklong Earth Day celebration.

In another instance, a group of biodiversity foundation members meeting in New Orleans desired a fact-finding tour of the southeastern Louisiana wetlands in order to witness areas of land loss. They asked to see the area around the bayou communities and to be escorted on a fishing vessel. My community partners and I made arrangements for a fisher, a resident, and me to accompany the group. During the trip, we told the visiting group about the causes and effects of coastal land loss. When we returned to the dock, the visitors commented that they would also need to confer with a scientist in order to find out the real causes for land loss, and proceeded to make arrangements for payment for the boat trip as if it were only a launch for hire. Several months later, when that same biodiversity organization was approached for information about pertinent foundation resources to address environmental degradation resulting from the BP oil spill, the foundation liaison felt the community and the area had little or no capacity to benefit from the foundation's level of expertise.

This contrasts with other cases when outsiders were introduced to the region by meeting community members in professional settings, and thus appreciated and understood their expertise. Those outside collaborators, after developing a relationship with the community, brought students, board members, or members of their organizational networks to the community to learn. In those instances, the community was engaged as a teaching entity with expertise, and when we interviewed the outside participants, they talked about their learning experience. Outsiders of this kind, and their organizations, continue to engage the community as experts. Community members have been asked for critique and commentary on coastal restoration projects, and have given testimonials in Washington DC.

Self-Governance

The Nobel Laureate Elinor Ostrom (2005) developed essential principles of effective self-governing, and we have applied these principles to our collaborative work between communities and researchers in coastal Louisiana. First, the boundaries of a common resource, and membership of the group that benefits from use and management of the resource, need to be clear. Then, there must be a balance between the effort that various members of the group put into managing the resource, and the benefit they each derive from the use of the resource. Overall, the benefits derived by the group need to be commensurate with sustainable yield from the resource they are managing. Members of the group need to have a say in making the rules that govern their use and management of the resource. They also need to have ways of changing rules that are not working effectively. Once management takes place, compliance of members with the rules that govern use and management of the resource needs to be monitored. The people doing the monitoring need to be accountable to the membership group, and members who break the rules need to be sanctioned, with the severity of sanctions matched to the seriousness of the infringement. Conflict will arise, thus the members of the group need to have access to low-cost mechanisms for conflict resolution. Outsider respect and the right of the community to manage are crucial. The sovereignty of the communities and the rights of the group to use and manage the resource need to be recognized by outsiders (Ostrom 2005).

The Community IRB or Community Declaration of Principles

One of the first required documents for an academic research team working with humans in the United States is an IRB, short for a no-harm statement that is presented to—and must be approved by—the Institutional Review Board from the home academic institution. I recommend that researchers and agencies create protocols with the community that go beyond the mandatory institutional IRBs when doing collaborative research in communities. These include respectful *entrée*, creating a mutual IRB with the community, and a possible declaration of the community's rights, all done in plain language and/or in the preferred language of the community (West et al. 2008, Peterson 2010). Documents created on the university or agency side also must be shared with the community so that there is full understanding of the purpose of the project and the work to be accomplished.

Community IRBs have been developed by overly researched communities and communities that have been used as subjects repeatedly without receiving benefit from the intrusion of time or from the results of the extracted data (West et al. 2008). The Declaration of Principles is the community version of an IRB. In the time following Hurricane Katrina, we developed several tools that would help define the communities' role in planning, research or collaboration with outside entities. Over the years, the dialogues we have had while developing the principles have helped the academic partners better learn from the community about their expertise, level of interest, and their capacity for participating in various stages of the work, as well as how the community wants to solve a problem using the knowledge gained. For the community, it has been an opportunity to better understand the dynamics of academic research, the academic disciplines involved, and the terminology and language of the academy. One community leader confessed that in previous projects, he often shook his head "yes" and agreed with the outside researcher even when he didn't understand what was being said. There can be many reasons for this—what might seem to be accommodation of the researcher—but from the community's perspective, it can be as simple as "I trusted the researcher so I said yes." In contrast, after working on several projects and participating in developing the community IRB and Declaration of Principles, one community member has said, "I have become multi-lingual. I can now speak my own language as well as being conversant in academic, agency and research lingo" (Community member, Lower Plaquemines, Louisiana, interview by author, 2008).

Communities that are facing many issues at once can often be a magnet for researchers and research teams. This deluge of research work can often add to the burdens the community already shoulders, and that can lead to diminished value of the research itself. The Declaration of Principles, since it is an adaptive—not static—document, can be a way to revisit pressures, timelines and outcome expectations throughout a project. In one instance, a nonprofit team that works with several Louisiana coastal communities asked the community partners about priorities and workload in an effort to discern if some work needed to be re-prioritized, and if there were other resources the community needed in order to meet current obligations. The communities decided that they would not take on any more research that didn't address some of their immediate issues or stressors, as they did not have sufficient time or the energy to use frivolously on results they likely would not benefit from.

The following is an example of a community IRB or Declaration of Principles that our community-researcher collaborative teams have developed. Again, the "Principles" are not a recipe; each setting and project is different. There are, however some values that remain consistent.

A Sample Community IRB or Community Declaration of Principles

- Names of Participating Entities: Agencies-Academics-Community
- Project Title

Declaration of Team Principles, Ethics and Expectations: This is a document to use as a mode of operation between agencies, academics and communities Declaration of Principles

- Community and [participating entities] and others:

These principles of Participatory Action Research (PAR) are based upon the underlying concept that such research is a dynamic process in which all participants possess the courage to engage in collaboration. Goals include balancing the historical and socialized power differential and promoting the self-actualization of collaborating parties. Essential to the success of such collaborations is a commitment by the large partner organizations to engage the community in ethical, equitable, respectful, and socially responsible ways, while making a conscious effort to *step away from the cultural hierarchy placing academic experience in a position of unequaled power.*

In order to best reflect PAR practices and to ensure constructive collaboration among the community and [the participating entities] and others, participants should strive to adhere to certain principles:

- Openness and Honesty

From the start, we strive to clearly explain the strengths and limitations of our participation, taking care not to make claims of results that exceed our true abilities. Such openness is intended to foster a true two-way relationship. We attempt to be transparent about our abilities and aims, and open to input and advice from *all participants*. Additionally, we will not harbor grudges; rather we will take the initiative to communicate our feelings about the PAR process and the information being shared.

- Clear Communication

Clear and inclusive communication among all participants is of the utmost importance throughout our collaboration. We intend to discuss the most effective forms of communication possibly integrating phone calls, emails, snail mail as well as visits. The intention is to facilitate communication between *all* involved parties, and to ensure that individual roles as well as institutional roles are as clearly understood as possible. Open dialogue is essential for all partners and information must be dispersed to *all* members participating in the process, by whatever means necessary. Due to the challenging nature of maintaining an open dialogue, it is necessary for collaborators to regularly assess the adequacy of dispersion of information to other members of the team.

- Commitment Against Harm to the Community

No harm should ever come to the community due to financial, intellectual, or other negotiations. If disagreements should occur between participants or organizations, it is imperative that the community's vision or wellbeing not be harmed. Results and information from this project should also never be used in such a way that benefits [participating entities] or individual academics without also benefiting the community, *to an equal extent.* Full disclosure of invocations of the community's name must be adhered to, in order to ensure that the exploitation of the community never takes place for the benefit of any others, *even in situations of innocent intentions.*

- Commitment to Resources

Within the context of limited resources, [the participating entities] are committed to helping the community with any available resources at their disposal and locating other partners in community enhancement and resilience.

- Valuing of Local Knowledge and Input

All team members must commit to a disownment of the oppressive structure of academic knowledge. We recognize the value and importance of the knowledge held by residents of the community regarding, but not limited to, physical, natural, and social environments. We fully intend to not only respect and consider such knowledge, but to consider it *equal,* if not more valuable, than academic/scientific knowledge. Local knowledge, reciprocal input, and a sharing of an understanding of each participating groups' respective values, and *equal intelligence,* are the foundations upon which this research is built.

- Inclusion of the Entire Community

We aim to be inclusive of all parts of the community. Through various methods of communication, we will make every attempt to ensure that we have some means to get in touch with all parties who wish to be involved in the research process. A very basic goal of PAR is to assist in developing personal/collaborative capacity among as many members of the participating organizations and community as is possible.

- Flexibility

We aim to be flexible in all of our collaboration, not only with the time and scheduling of meetings, but also with the approach of the work we are collaborating to do. In PAR, deliverables should be couched in broader terms. Such an approach will allow the collaboration to be more flexible. The goal is to be adaptive, so that a change in course can be facilitated if such a need should arise or be perceived as beneficial.

- Consideration of Time

We recognize that the community is busy, and that time spent in collaboration is time spent away from family, work, leisure or other activities devoted to community resiliency. Research and other activities must be conducted in a manner that complies with the time schedule and needs of the community. The

PAR process requires a significant time investment from all partners. The forming of relationships and building of trust are not processes that should be rushed. Participants must be aware that the amount of work required in such a project will often go outside of the parameters of the common workday, and time spent in collaboration will more often than not exceed that which was anticipated.

- Placing the Vision of the Community First

Regardless of what [the participating entities] or academic partners aim to achieve from the standpoint of applied research, the highest priority will always be placed upon the needs and concerns of the community. We recognize that the community's willingness to collaborate with us is a privilege, and not a right. That being the case, we will do everything in our power to continually earn the privilege of working with the community, in part by never letting our concerns impede the community's vision.

- Reflection

Ongoing participatory evaluation is critical, and must be given priority in discussions and communications. The PAR process is one of nearly constant reflection, and we are committed to continual and collaborative evaluation of whatever course the research takes. This is done in order to ensure that best collaborative practices are being followed and that team members are consistently asking, and addressing, the questions that are most relevant to the community.

- Sharing of Research

We recognize the potential value that the project's research may have for all participants once it is completed. We intend to fully and equitably disclose all information and results and collaborate in preparing and editing all research documents. All team members will have *equal* stakes in the final product and thus, equal input and veto power regarding the sharing of results or conclusions outside the team.

- Rights of the Community

We, the people of the Louisiana coast, are the most historied region of North America. Because of our resiliency, efficacy, capacity, innovation and stewardship of our home lands some Louisiana coastal communities have existed for hundreds of years. Therefore, we retain a long-standing and immediate expertise of our estuaries, bayous, and coastal regions.

Due to the recent history of hurricanes and the present oil disaster many people want to be present in our communities impacted by disasters. We invite equitable and reciprocal partnerships and nothing less. In so doing we declare the following rights of our community.

Persons, who together form a community, local community, original local community shall enjoy the rights to conserve or restore customs and tradition, local knowledge, arts and culture of the locality and to participate in managing, maintaining, deriving benefits from the natural resources and the environment in a balanced and sustainable manner.

As a person/organization/agency who wishes to interact with the community the following community rights will be observed.

1. The right to transparency in all information regarding the purpose of the visit and other information including:
   - Who is the visitor representing
   - Reasons or motives for extracting information
   - How the data will be used
   - How identity will be protected
   - Will there be compensation for time and expenses, if so how
   - What method was used to obtain permission
   - Who has final authority in the determination of how and how long the data will be used
2. The right to forbid the use or publication of any kind of communication and/or quotation without expressed written consent of said resident. No communications or quotations shall be used into perpetuity without expressed written consent.
3. The right to forbid the use or publication of images of citizens and residents and/or their property without expressed written consent. No images or data shall be used into perpetuity without expressed written consent.
4. The right to retain ownership of any information extracted from data gathered through observation and or participation with the community and its residents.
5. The right to decline participation in any capacity regarding the purpose of the visit.
6. The right to be left alone.
7. The right to say no to the visit or research.

Conclusion

The health and wellbeing of a community or participating group depends on using appropriate and respectful methods in every phase of a project, be it research or applied work, engaging with the community from project conception to implementation, and in monitoring and evaluation of the work. Truly participatory work results in co-learning by all parties and offers a venue for building long term relationships as well as a foundation of data and resources on which future work can be based. In regions that are as disaster prone, as in coastal Louisiana, building relationships with resident communities is essential for being able to quickly respond to perturbation, as well as for collecting data critical for long-term work and research. Participatory methods allow for boundary crossing that enables academics, government agencies, and communities to solve problems and improve lives. The field of participatory work and co-management has grown over the past 10 years due to demands for federal agencies and foundations to include the participation of community members. It is the responsibility of the outsider to use and abide by appropriate tools of engagement to achieve respectful and efficacious collaboration.

References

Blackburn, J., ed. 1998. Who Changes: Institutionalizing Participation in Development. London: ITP.

Brasseaux, C.A. 2005. French, Cajun, Creole, Houma: A Primer on Francophone Louisiana. LSU Press.

Cooke, B., and Uma Kothari. 2001. *Participation: The New Tyranny*. London: Zed Books.

Denzin, Norman K., Yvonna Lincoln, and Linda Tuhiwai Smith. (2008). *Handbook of Critical and Indigenous Methodologies*. Thousand Oaks: Sage Pub.

Fals-Borda, Orlando, and Mohammad Anisur Rahman. 1991. Action and Knowledge: Breaking the Monopoly with Participatory Action-Research. MIT Press.

Fischer, Frank. 2000. Citizens, Experts, and the Environment: The Politics of Local Knowledge. Durham: Duke University Press.

Forester, John. 1989. *Planning in the Face of Power*. Berkeley, California: University of California Press.

Freire, Paulo. 1974. *Education for Critical Consciousness*. London: Continuum.

Freudenburg, W.R., R. Gramling, S. Laska, and K.T. Erikson. (2009). Disproportionality and Disaster: Hurricane Katrina and the Mississippi River-Gulf Outlet. *Social Science Quarterly*, 90(3): 497–515.

Gaventa, John. 1993. The Powerful, the Powerless, and the Experts: the Knowledge Struggle in the Information Age. In *Voices of Change; Participatory Research in the United States and Canada*, ed. Peter Park, 21–40. Westport: Bergin & Garvey.

Habermas, Jurgen. 1991. *The Structural Transformation of the Public Sphere: an Inquiry into a Category of Bourgeois Society*. Cambridge: MIT Press.

Jervis, Rick. 2014. Louisiana Bays and Bayous Vanish from Nautical Maps. *USA Today Online*, February 12. http://www.usatoday.com/story/news/nation/2014/02/12/noaa-maps-disappear-coastal-erosion/5259611/.

Kunde, James. Presentation at the American Meteorological Society. January, 2007. San Antonio, TX.

List, J.H., A.H. Sallenger Jr, M.E. Hansen, and B.E. Jaffe. 1997. Accelerated relative sea-level rise and rapid coastal erosion: testing a causal relationship for the Louisiana barrier islands. *Marine Geology*, 140(3): 347–65.

Ostrom, Elinor. 2005. *Understanding Institutional Diversity*. Princeton: Princeton University Press.

Park, Peter, ed. 1993. *Voices of Change; Participatory Research in the United States and Canada*. Westport: Bergin & Garvey.

Peterson, Kristina. 2010. *Transforming researchers and practitioners: The unanticipated consequences (significance) of Participatory Action Research (PAR)*. PhD Dissertation, University of New Orleans. New Orleans.

Rotkin-Ellman, M., K.K. Wong, and G.M. Solomon. 2012. Seafood contamination after the BP Gulf oil spill and risks to vulnerable populations: a critique of the FDA risk assessment. *Environmental Health Perspectives*, 120(2): 157.

Smith, Linda. 1999. *Decolonizing Methodologies: Research and Indigenous Peoples*. London: Zed.

West, J., K.M. Peterson, M. Alcina, and S. Laska. 2008. Principles of Participation and Issues of Entry for Participatory Action Research (PAR) in Coastal Community Resiliency Enhancement Collaboration. *Journal for Community Engaged Research and Learning Partnerships*, 1(1).

Chapter 7

"We Can't Give Up":
A Conversation About
Community Engagement

Albert P. Naquin, Amy E. Lesen, and Kristina J. Peterson

Albert P. Naquin is Chief of the Isle de Jean Charles Band of Biloxi-Chitimacha-Choctaw Indians.

The following paragraph is adapted from the Isle de Jean Charles community's website (http://www.isledejeancharles.com).

Isle de Jean Charles is a narrow island deep in the bayous of South Louisiana, approximately 80 miles (126 km) from New Orleans. A place of immense physical beauty and great biodiversity, it is most importantly home to our Native American community, The Isle de Jean Charles Band of Biloxi-Chitimacha-Choctaw Indians. Our language is French and our way of life is simple but abundant. For the people of Isle de Jean Charles, the island is more than simply a place to live. It is the epicenter of our people and traditions. It is where our ancestors cultivated what has become a unique part of Louisiana culture. Today, the land that has sustained us for generations is vanishing before our eyes. Our tribal lands are plagued with a host of environmental problems—coastal erosion, lack of soil renewal, oil company and government canals, and a rising sea level—which are threatening our way of life, our culture and our identity on what has become our gradually shrinking island. The Louisiana wetlands are home to some of the world's most unique and diverse cultures and communities. Few communities have been affected by wetlands loss as directly as Isle de Jean Charles. After numerous hurricanes, the Isle is a fragment of what it once was. The beautiful Willows and Live Oaks that once lined Bayou Jean Charles, and were the childhood playground for the children of the Isle, are a thing of the past. More than a few houses maintain damage from storms gone by. The 2010 BP disaster and cleanup efforts resulted in exposing members of the Isle de Jean Charles community to chemicals, leaving them at risk to long term health effects, and the waters and the bounty they historically provided Tribal members have been forever changed. Fish with tumors, and oyster beds destroyed, have had a drastic effect not only on income, but on the ability to be a self-sufficient people. Saltwater intrusion and land loss have made the tradition of growing our own produce a thing of the past. The cattle that were once plentiful, no longer have adequate grazing grounds. The herbs used to treat the sick, are a thing of the past. The result, Tribal members struggle with the increased cost of living from having to shop for the food they once readily harvested. Modern diseases linked to nutrition are on the increase and members now have to turn to modern medicine to cure them. Chief Albert and his community are also currently working to gain status from the United States Bureau of Indian Affairs as a federally recognized tribe, which would vastly increase the resources available to the community.

Kristina Peterson first became acquainted with the tribe after Hurricane Andrew and helped develop a response group for the region to give assistance to rebuilding (TRAC). Years later she visited the Island after Hurricanes Isidore and Lili in 2002. She was shocked by the amount of land loss. Her attempts to secure national funds for the Island and surrounding area were unsuccessful. Her work with the Island began in earnest after Hurricanes Katrina and Rita and has continued since.

Peterson introduced Lesen to Chief Albert and members of the neighboring Pointe-au-Chien community in 2010. Since then, Lesen has spent time getting to know Chief Naquin and learning about his community's challenges.

The following is the edited transcript of a conversation between Amy Lesen (AL), Chief Albert P. Naquin (APN), and Kristina Peterson (KP), on January 3, 2014 in Montegut, Louisiana. Amy Lesen recorded, and transcribed the interview, and edited the transcription in collaboration with Naquin and Peterson.

AL: What has it been like for you to be approached by people? Has it always felt like people come to you with a certain intention—do they have a spoken intention versus and unspoken intention? Are people invited by you, to work with you, or are they uninvited? Kris mentioned when we were talking earlier today that with some people it's kind of like a "drive-by:" they will come and they'll get information from you and then go away. So, can you talk about the different ways that has been for you?

APN: Yeah, until we met Kris we had a lot of people who would come by—like a drive-by shooting, you called it. I guess it started more after [Hurricanes] Katrina and Rita ... because of New Orleans, probably. You know, when we get affected here, we get a few visitors here and there, but mostly it was after that. We met a lot of people who wanted to come and help us to do certain projects. One thing we went though was outsiders that kind of spoke for the community, which was unfair for us. Then we've had folks who would come and do pictures, and interviews, just to try to get us some money or some aid in some kind of way, especially after disasters, that's when you see a lot of people. We had many, many, videos but also newspaper [journalists], they would come, and some even wrote books. But nothing would ever come back to the community. They would come and write a book and write stories about what the people went through, but then nothing would come back to the community. They might give us a book, which is fine.

I guess it was in '05 after [Hurricanes] Katrina and Rita we had a crowd underneath my brother's house and we had two people there, a man and a woman. They said they had contacts and could get us a Bobcat [the manufacturer's name for a type of soil excavator–ed.] and a forklift and the trailers that we needed to haul them. But it never happened. We haven't seen them since then, we haven't heard from them.

AL: And do you know where they were from? When people find you this way, do they communicate who they are, why they're there, what their intentions are, how they found you?

APN: They were supposedly some non-profit people, the lady was from California, the guy I don't know where he was from. How they found us, I really don't know. We never asked that question. I think they gave us a card that had their names, but they said they was gonna contact us as soon as they had the stuff for us, but it never happened. And other [organizations], after Katrina and Rita, they wanted to do certain things and I said, "But that's not what we need." They said, "You're not the only community we work with." I said, "I understand that but this is what we need for the community, and the other communities need something else." So I made a couple of meetings and then I walked away. One of the reasons we don't like working with them, every time we go to a meeting they say, "Give more money to the non-profit." But I'm here! They say, "We help the communities." But I say, "We're here!"

Then we met Kris and Kris has never let us down yet, she's been one of our champions helping us along. She always brings us to talk. She probably knows enough about the community to do a good job at it, probably better than I can, but nevertheless she brings me so they see me, they can hear me. That's the good part about it. So we've had a whole bunch of them, but ... she hasn't given up on us. She takes probably a lot of criticism but she hasn't given up on us.

AL: What do you feel like is the big difference between your experiences with people who have done these drive-bys? Sometimes people have taken resources that you could have used, or were trying to speak for you. What is it about the way Kris works that is so different?

APN: She don't let up! We have others that come maybe once, maybe twice, I think it's mostly just to get information, nothing serious for the community to benefit. And the good part about it, she knows how to handle the money, we need a big brother to get a grant and, we do have 501(c)(3) [US tax-exempt status for NGOs–ed.] but we need somebody to school us on what we gotta do to get some more. So with Kris and the people she works with, we do have people who care and are serious and sincere about trying to help the communities instead of just trying to get information.

AL: So it sounds like in terms of a commitment it makes a really big difference to have someone who is around consistently. Is it about personal relationships being important? It sounds like the presence and commitment are really important.

APN: Oh yeah, well, Kris has been around us long enough to get to know us. To me, I guess, when you're committed, you're serious. Like Kris sends us places where we can get well known. Because [otherwise] people just don't know us.

AL: So it sounds like it's the commitment. Tell me if I'm right or wrong in what I heard: that someone who speaks for you, versus someone who brings you to speak for yourself, is important?

APN: Yes, yes. It's best, I guess, because I lived it. I lived it and I seen how [Isle de Jean Charles] was when it was paradise. Now a lot of people say it's so beautiful but they didn't see it when it was beautiful, all the trees, when we had the cattle and the chickens, we had horses. Basically we had a nice little farm place. It wasn't big but people had their livestock, their gardens. So I seen that. When you start talking about things that happened so long ago, like Wounded Knee [a massacre of Native Americans that occurred in 1890]. I didn't [always] know that happened in the 1800s. But it was. Even my grandpa was born at that time. I wasn't that far off, it almost happened in my time. The Island is not that old, but when it first started with the people moving there, they could walk from Montegut, they could walk from Pointe-au-Chien, and go all the way to the lake on foot. That's why there's a place we call "the end of the Island" because you could walk there. That's where my grandpa settled at.

AL: So what I think I'm hearing you say, too, is that part of what you have to say is what your story is, and nobody else can tell your story.

APN: Yeah. I remember.

AL: Another thing that is interesting is that there are all these different categories of people you've worked with: journalists, academics, people in government, people in non-profit. Are those categories important to you when you're working with someone? Is it important to you if they are a journalist, or an academic, or do they even communicate that to you?

APN: At one time it was important to me because I wanted to get the word out. But the word goes out and nothing happens. Journalists doesn't do me anything. People do tell us who they are and they might even give us their card. I guess it's their living: they want to write their story. They get good stories and good pictures. So they are doing their job, but then coming back to get the job done that we are looking for, it doesn't happen. They get theirs but we don't get ours. It's almost like paddling upriver, you keep paddling but the river is too strong. I even took some journalists to the cemetery down Pointe-au-Chien, because the cemetery is going underwater, so I brought them there and every place they want to see. The stories they wanted to hear, the questions they would ask, we gave good answers. So they got their good stories but we never got anything in return for it.

AL: How is it different when someone comes from a non-profit organization?

APN: The non-profit people come and they say "We want to help with this or that." They tell us how they want to help us. Some of the non-profits have said, "We can talk for you." We say, "Well, we can talk for ourselves." They seem to have money for their travel and salary but to me the non-profit is just a job for them. So we kind of get discouraged.

AL: I know another category of people you've worked with is academics. You told me that you've worked with anthropologists, scientists. How has that been for you, and how has that been different than working with journalists or people from non-profits? What has it been like when someone says to you, "I'm from a university?"

APN: Them I don't mind, when someone is going to do a study or I suppose do something great in their career. For example through e-mail, a woman from Europe told me she wanted to write about a small community like ours. So I sent her a lot of information that we had, the history. She put that together and I told her, "I hope you get an "A" on that thing." (Laughs). And there was another woman like that from Pennsylvania. I suppose that was different because I had done it before and I was prepared. Sometimes they don't have enough money, I suppose, to come here and take the pictures and interviews so we do it by e-mail. But some, for example a woman who is I think an anthropologist who is working on her PhD ...

KP: Yes, she is an anthropologist.

APN: Well, she did a lot of digging, and she helped us a lot, she showed us how to go dig for stuff [find information and history–ed.]. Then we had another girl who came and she found plants we didn't even know we had. So we had quite a few people who came down here that was good for us. Let me tell you, the students are better than the journalists!

AL: Tell me why.

APN: They come here, they dig for stuff that we are digging for also. Like for the plants, we thought we knew them all. But [that one graduate student] came around and found stuff, I guess she looked at books and then went and looked around and found [plants] we didn't even know we had. Probably the elders had used it. But on the Island, it's tough. We have the *fraises*, and the cactus ... we used to clean it and boil that and drink that. That's all we have left.

AL: So it sounds like sometimes the students from universities, their research is actually helpful to you?

APN: Yes, they help us by whatever they are doing, and they always give us a copy of whatever they wrote. Like for example [one researcher], she wrote a chapter all about the community down here.

AL: So it sounds like sometimes people come from a university and what they want to do is also helpful for you. Do people from a university ever come to you and ask you what they could do that might help you?

APN: Yes, they did. They have said, "You are helping me so what I can I do in return to help you?" Most of the time if they are students we try not to put them to work. (Laughs). We had a couple of them come do that.

AL: Do you think when researchers come and ask you that, do you think there is something they could do to help?

APN: They could. If they are doing research where they could maybe find us some real Indians to trace back to. Because it takes a lot of money and time for us to dig for information. We could tell people to look for certain names and dates for us. But that hasn't happened yet. There is the one graduate student who has tried to find some information for us, she is a real go-getter.

AL: What about scientists?

APN: Well, I guess scientists are so smart they are dangerous.

AL: Why?

APN: I guess they can't come up with a definite answer. There are too many "ifs" and "ands." OK, compared to myself: I am very simple. I know a car runs but I don't need to know why. But a scientists needs to know every little thing about everything, about whatever they are scientists of. If it's plants or what have you, they want to know everything, and that can be kind of confusing to me. Like electricity. I can wire a house and I need to know the right sized wire, but there's all the other little numbers in there like the ohms, and the voltage and the watts, I don't need to know that. I think the scientists are good and we need them, and I guess if you put enough of them together, they will come up with an answer. (Laughs). I say that because once at a workshop [for the North American Free Trade Agreement–ed.] in New Orleans, I was sitting at a table with three scientists and some people from a non-profit. They came up with the solution and they just kept discussing it. One would say something and the other one would say "But what if ..." And then they wanted me to give the presentation and I said, "No way!" At the end of the resilience meeting last year, they asked me "What would you do?" I would tell the scientists to tell somebody *else* to come and tell *me*, so I could understand! (Laughs). Because the scientists, they don't say enough. They know so much that they just leave me by myself. So if the scientists could explain to someone who understands, but can come down to our level to explain it to us.

AL: You mean like a translator?

APN: Translator, yes.

AL: This is what I'm hearing you say, please correct me if it's wrong: I heard you say it's clear that scientists are doing work that's important.

APN: Yes.

AL: And maybe even important to your life.

APN: Yes.

AL: But the problem is?

APN: To understand them.

AL: Communication.

APN: Yes, communication. Yes, that's basically what it is.

AL: Why is that? Do they want to? Do they not want to? Are they not capable of explaining? Have you thought about that?

APN: I haven't thought about it, but I can give you my honest opinion. I think probably they could but they are so used to talking to people in their job, so they all understand each other. So when they come to talk to people like us, average people, they kind of forget, I guess, and they explain it the way they explain to their coworkers. Somebody like me, I understand what they are trying to say, but it's a matter of connecting the lines.

KP: Have you thought about it from the standpoint of the folks who speak French and the folks who don't speak French, only English, and they have no clue of what you're talking about? The scientists have one language and maybe that language is all they can communicate with, and it's not that they are smarter than you, but maybe they have an inability to speak another language?

APN: That makes sense.

KP: But it seems like you have more ability because you are multilingual, you are at least picking up some of the things that they are talking about, and some of the other languages from some of the agencies. Whereas they have not tried to speak your language. So, you are much more multilingual than they are. So, who is the intelligent one here?

APN: Hmmm, yeah, I understand.

AL: You said that one of the reasons you like working with Kris is because of her commitment to spend time with you in your community, that she is part of your community almost.

APN: Yes, she is.

AL: And so she has made this commitment to get to know you? You used those words.

APN: Uh huh.

AL: So, hypothetically, if one of these scientists did the same, could that be a solution? You said earlier you think scientists could figure out how to explain, but they aren't used to it because they spend all their time with other scientists. If a scientist came to spend some time with you and get to know you … ?

APN: Maybe I could teach them something! Teach them what's happening, maybe they could come sit down in the real world, and figure out why this occurs with the plants or the fish, then they could see it and feel it and get dirty in it. Yeah, I think I could learn a little bit about what they do and they could learn about us and it would be beneficial for scientists and the local folks, I think it would be very beneficial.

AL: So, it sounds like a lot of people come to you, maybe invited, maybe not, but you haven't always gotten the outcome you wanted?

APN: Yes.

AL: What would be helpful to you? What is the way it would work better for you?

APN: You want to hear my big picture? My big picture is to have everyone together again. There was no such thing as a locked door. We used to have a little wooden thing for our door so the wind wouldn't open it, to keep the dogs and the chickens out, just a little latch. I know we couldn't go back to that exactly, because there are too many people. But to have a community center, my vision is to have what we used to do on the Island. At three-o-clock everyone would gather at someone's house or at the store. They didn't used to drink coffee, they used to drink beer, but it was like tea time in England, we would have some fry bread and coffee. First we need funds and land, it would be an ideal thing to start all over again, to try to get our kids and grandkids to get the culture back. Because, probably after my generation is gone, we won't have our culture anymore because those are dying out pretty quick. We would probably have to have people from out of the state or the community to teach us things. But to have everyone get together to do our crafts.

AL: How could someone from a university … if someone has intention to be helpful, what could they do to help you get closer to your perfect world?

APN: Start preaching like I do. To tell people how we used to live versus how we live today. I know no one wants to go back totally to living out in the cold, go hunting when you don't want to, or wash our clothes in the bayou. I know no one wants to go back to that. But we want to stay with the modern world as a modern community and [also] do some of what we used to do, a little gardening, a little farming, because that is part of our culture. To have a place or time to do our cultural work.

AL: So what do you mean when you said people could start preaching like you do?

APN: To tell our story. To say this group here used to be a family community that Mother Nature washed away and the people had to move off, and there is no more gathering.

AL: Help you tell your story?

APN: Yeah, help us tell our story.

KP: To put it another way, you are wanting to be in your community in a very subsistence, a very sustainable way. Are you inferring that any way scientists could help you get you back to your sustainable self-sufficiency is what would be helpful?

APN: Yes, that makes sense.

AL: But no scientist has ever gotten there? I think you already said why. You said earlier that no scientist has ever spent time with you to find out what was important to you?

APN: That sounds exactly right. What's important to them is different than what is important to me. Somebody has to break the ground, like a groundbreaking.

AL: What is interesting to me is that when I ask you what your perfect world looks like, and I ask you about scientists, it didn't even really occur to you that an academic or a scientist could even help you with that. Because it's never happened?

APN: No, it's never happened.

AL: It sounds like there is such a big agenda difference.

APN: I guess if you take the scientists and you take us and other people in between me and the scientists, and put us all together, you could make a jambalaya! And probably that's what needs to happen. Put different people in the same pot, to try to accomplish that goal.

AL: Can you talk a little bit about your role in your community and what you and your community are trying to accomplish?

APN: I guess I'm like the president, like the governor. With some other people helping me out. I'm not alone. But my role is to see what the community needs right now. Without the community, I'm nothing. If our community dies, well, then I die with them. So I guess my role is to make sure that we don't die, that we just keep rolling. My grandpa seen the community almost like a flower blooming, and now mine is just going back in. So, he seen it grow and I'm seeing it die. My biggest challenge is to make sure that we can get a new [home for the] community so that we can survive. So my role right now is to try to get a new community, and to make sure the people are happy and that they are taken care of in certain ways, and to get them to a point where they won't need a handout, where we can be self-sufficient. That to me is very important, to get people to be self-sufficient. As long as we live on the Island it won't happen. So I guess my role is to better each household, I guess, to where they can benefit from what they do every day. My job is to make sure that happens, and that's the hard part. Hopefully if we get a new community, it will all be in their backyard. If we have crawfish, if we have cows, some type of farming, it could be in our backyards, we wouldn't have to go far for that. My goal is to get the people to be self-sufficient, and my role is to accomplish that goal, and that goal is tough.

KP: You talk about being witness and it seems like that's also part of your role, is to be that witness for your community and your place.

APN: Yeah.

KP: And in some ways that's a very difficult role, but you embody that.

APN: I'm glad to hear that.

KP: And one other thing you just touched on now is that no matter how good people are from the outside who want to help, it still takes resources from you, your time that you could spend doing other things. Do you ever get tired of that? Is that in some ways energizing or is it tiring? How you're almost constantly entertaining people from outside.

APN: Yes, I guess I do get tired. At the beginning I didn't because I was fighting to reach my goal. But then it doesn't happen, so then you have to back off and the excitement is not there anymore. It's not that you don't want to do it, but it serves no purpose. So, I guess you just lose your energy. I get tired and lose interest because I'm doing it for nothing. Actually there is somebody who wants to come visit now, who wants to come down, she's from Australia, who wants to ride a boat. So I guess I have to find her a boat.

AL: What I hear you saying, if I'm correct, is that you would have the energy to spend to do all these things people from the outside are always asking you to do, if you really thought it was going to help. And it sounds like there's been so much time and energy spent, and you haven't seen results, is that what makes you tired?

APN: Yeah.

AL: It was also interesting earlier when I was asking you questions about academics and people bringing students and you kept saying, "I didn't mind!" But saying "they came and I didn't mind," is so different from saying "they came and it was so helpful!"

APN: (Laughs). Yes, I guess that's true!

AL: You know the saying, "a picture speaks a thousand words?" To me that said a lot. The best thing you could say about many of your experiences was "I didn't mind."

APN: Yeah, there's no excitement, huh? But it's kind of hard to give up on people because you just don't know when that right person is going to show up. It's like when you dig for gold, you have to dig through the sediments and shake it out. You never know when that big gold nugget is going to show up. Maybe it's not exciting, but then when that person comes along, that would put some excitement in you. So you don't want to throw too many of the journalists or other people away because you don't know when one is really going to catch on fire.

AL: It sounds like you've spent a lot of time sifting for those nuggets.

APN: Yes, but we can't give up.

Chapter 8

A New Paradigm For Science Communication? Social Media, Twitter, Science, and Public Engagement: A Literature Review

Amy E. Lesen

Scientists who engage with society perform better academically #scicomm … [1]

Retweet if you agree: It's time for Congress to stop denying the science on #climatechange.[2]

Ocean acidification—one of the "top 5 ocean stories of 2013" according to Smithsonian Magazine …[3]

No honorable scientist would say "ocean acidification" because it is a shibboleth, a big evil climate lie designed to deceive & stampede.[4]

Civic engagement activities raised the odds of graduation and improved high school students' progress in reading, math, science and history.[5]

Looking to see how many teachers/educators are using twitter. RT if you are an educator using #twitter! #edchat #education …[6]

Social media and other web-based and mobile technologies are transforming communication. In the field of academic biophysical science, this phenomenon is creating a rapidly shifting landscape of science communication and civic

---

1 Tweeted September 26, 2012 by @scientistmags (this and all other quoted tweets accessed via Topsy.com).
2 Tweeted May 8, 2013 from Barack Obama's official Twitter account, @BarackObama.
3 Tweeted December 31, 2013 by @oaicc_project.
4 Tweeted December 31, 2013 by @@donbeeman.
5 Tweeted May 22, 2012 by @citizenshipcnts.
6 Tweeted April 2, 2013 by @slackt.

engagement containing new spaces, means, and possibilities for information exchange, knowledge dissemination, dialogue, and citizen empowerment. The social media platform Twitter is playing a significant role in that changing landscape (Federer 2013). Twitter can be used at every step of the process of scientific research from inception to dissemination, and in some sectors of the academy this is being encouraged to such an extent that detailed manuals are now available for using Twitter in science (Mollett, Moran, and Dunleavy 2011). Scientists can now communicate with their peers across the globe more easily and quickly than ever before, and non-scientists can read and participate in these Twitter exchanges as well. Twitter, social media, and mobile applications are being used as tools to innovate "citizen science," giving the lay public new ways to collect data and participate in scientific research. Many scholars have suggested that academe must now respond to these shifts and support scientific public outreach through social media.

What does this mean for science and scientists? Does having access to the daily ponderings of a university chemist in her laboratory give us new insight into her humanity? Will this allow her to gain our trust? Does giving the public unfettered access to information strengthen democracy? Does this, in turn, also upset traditional notions of expertise and power? In this chapter, I will reflect on the implications of these and other questions. I present an overview of some recent literature and online writings about science, social media, and Twitter, along with data gathered from Twitter's database, to consider how this new communication paradigm is creating change in a number of spheres: the practice of science, the way academic institutions function, the role of scientists in the public arena, our conception of expertise, and the dynamics within and between the scientific and lay communities.

Definitions and Methods

Many of the terms I will discuss here have multiple meanings depending on the context in which they are used, so it will be helpful to define these terms at the outset. Burns, O'Connor, and Stocklmayer (2003) give useful definitions for most of these terms in their article, *Science communication: a contemporary definition*, and I will rely heavily on that work. I use the term "science" here to mean "mathematics, statistics, engineering, technology, medicine, and related fields" and the term "scientist" to mean a practitioner of one or more of those fields "in industry, the academic community and government" (184–5). This is in contrast to the way I use the terms "social science" and "social scientist" to mean the fields and practitioners of sociology, psychology, economics, anthropology, political science and other fields of the study of human interaction and organization (Taylor Baines & Associates 2006). Many scholars prefer to use the contrasting terms "biophysical sciences" and "social sciences." My use of the terms "science" and "scientist" here is not meant to suggest that the social sciences do not employ quantitative methods or are not sciences in and of themselves. Rather, my usage here reflects the fact that the subjects of this article are the biophysical sciences and biophysical scientists (not the social sciences and social scientists) and thus it seems unnecessary to repeatedly use the "biophysical" modifier.

I will use the term "the public" here to mean anyone who is not a practitioner of science. Burns, O'Connor, and Stocklmayer (184) rightfully point out that "the simplest and most useful definition of the public is every person in society," a definition that, of course, includes scientists. But because a main purpose of this chapter is to scrutinize interactions between scientists and non-scientists, it will be useful here to draw a distinction between scientists and other members of the public.

I will adopt Burns, O'Connor, and Stocklmayer's (191) definition of "science communication" as:

> the use of appropriate skills, media, activities, and dialogue to produce one or more of the following personal responses to science ... Awareness, including familiarity with new aspects of science; Enjoyment or other affective responses, e.g. appreciating science as entertainment or art; Interest, as evidenced by voluntary involvement with science or its communication; Opinions, the forming, reforming, or confirming of science-related attitudes; Understanding of science, its content, processes, and social factors. Science communication may involve science practitioners, mediators, and other members of the general public, either peer-to-peer or between groups.

It is important to note that science communication includes communication between scientists, as I will discuss peer-to-peer science communication extensively in this chapter.

Ehrlich (2000, vi) describes "civic engagement" as "working to make a difference in the civic life of our communities and developing the combination of knowledge, skills, values and motivation to make that difference. It means promoting the quality of life in a community, through both political and non-political processes." For Gordon, Baldwin-Philippi, and Balestra (2013, 2), civic engagement encompasses:

> ... three major categories comprising the ability to (1) acquire and process information relevant to formulating opinions about civic matters, (2) voice and debate opinions and beliefs related to civic life within communities or publics, and (3) take action in concert and/or tension with social institutions such as political parties, government, corporations, or community groups.

The specific definition of "public engagement" I will use here is slightly modified from that of Poliakoff and Webb (2007, 244): any communication or collaborative activity that engages science practitioners with non-practitioners of science.

In addition to searching the scholarly literature, I also used Google searches of the World Wide Web while researching for this chapter. Not surprisingly, a great deal of discussion and analysis of science and Twitter occurs on web-based blogs and in other online writing platforms which would not appear in a search for peer-reviewed scholarly publications, so it was important to those Web sources here. I also collected Twitter data directly. Twitter has what is known as an "open API" (application programming interface), enabling other web platforms to gather information from Twitter and making it possible to use Twitter itself as a research tool (Cook 2011; Federer 2013; McCormick et al. 2013; Rogers, Stevenson, and Weltevrede 2009). I used the paid, web-based service Topsy.com to collect all the raw data about Twitter usage I report here, unless otherwise cited.

Web 2.0, Social Media, and Twitter

Social media is part of the wider virtual space known as "Web 2.0." For Anderson et al. (2009, 1), "'Web 2.0' is an umbrella term that is used to refer to a new era of Web-enabled applications that are built around user-generated or user-manipulated content, such as wikis, blogs, podcasts, and social networking sites." For my purposes, social media include any Web 2.0 platform whose express purpose is to foster social interactions, exchanges, and networks. Some of the most well-known and widely used social media platforms are Facebook, Twitter, and YouTube, and some consider blogs to be a form of social media as well (Fullick 2012; Obar, Zube, and Lamp 2012). My focus here is Twitter because it seems to be the most popular social media platform amongst practitioners of science (Federer 2013; Letierce et al. 2010; Vence 2013).

Twitter was founded in 2006 in San Francisco, CA. The first Twitter update ("tweet") was posted by Jack Dorsey, one of Twitter's founders, on 21 March 2006. Twitter's platform was publicly available by November 2006 (Arceneaux and Weiss 2010; Johnson and Yang 2009; Murthy 2011; Topsy 2013). Tweets are limited to 140 characters in length, as are short message service (SMS) or text messages. This enables a great deal of flexibility and mobility in how users send updates to Twitter: tweets can be posted via text-messages sent from a mobile phone, one can also post via a large number of mobile phone applications ("apps"), or from the Twitter website (Johnson, and Yang 2009). Social networks on Twitter are set up by "following" a Twitter account—essentially by subscribing to another user's updates. Twitter nomenclature includes an @ at the beginning of the account name, thus @NASA is the official Twitter account for the US National Aeronautics and Space Administration, and @MichaelEMann is the Twitter account of the climate scientist Michael E. Mann. Hashtags are words or phrases preceded by the hash (#) symbol that can be invented or used by any Twitter user at any time. It's possible to search for all the tweets that contain a particular hashtag, enabling a user to follow any topic on Twitter with one or more hashtags associated with it. For example, if I post a tweet that includes the hashtag #sciencecommunication, other Twitter users will see that tweet in a search for that hashtag.

Twitter is relatively unique among social media platforms in that updates can be a one-way process: the person/account being followed is not obligated to follow. For example, I may have 200 followers (who can all read my tweets), but I may follow 400 people (whose tweets I can read), and neither of those groups need share any people in common. Tweets are often not directed to any particular Twitter account, although it is possible to direct-tweet another user or reply to a tweet. Twitter is often called "micro-blogging:" tweets, like blogs, are broadcast and available to anyone who chooses to follow them. A tweet can consist solely of a short message of 140 characters, or it can contain links to any longer content that has a web address. There are users on Twitter with just a handful of followers, while as of March 2014, Barack Obama has 42.3 million followers (Twitter 2014). Twitter can thus be a hugely effective way to communicate with a large group of people anywhere in the world.

Emerging Themes 2008–2009: "Bench-to-Bedside" or "Substantially Asocial?"

It wasn't until 2008–2009 that we began to see increased traffic on Twitter about scientific topics. The hashtag #climatechange, for example, had fewer than 10 mentions in any one month period on Twitter worldwide until December 2008, and didn't reach over 100 monthly mentions until March 2009 (most of the Twitter data I present here is reported by month). A broader search reveals that the first month during which the word "scientist" received more than 10,000 mentions was April 2009 (17,222 mentions). Analysis and discussion about science and Twitter began reaching a substantial volume during these same years, in the scholarly literature and on the Web. During this early period of science on Twitter, it quickly became clear that Twitter was a powerful tool both for peer-to-peer science communication, and for scientific public and civic engagement. Twitter had potential to significantly impact the practice of science, but also the way scientific information was disseminated and the impact science could have in society and the civic realm. This is illustrated well in an article about social media and health science from a conference proceeding from autumn 2009:

> Online and social media are practical tools for supporting distance collaborations relatively inexpensively while offering the added benefit of placing selected information in online spaces that facilitate discovery and discussion with clinical care providers, thus supporting the fundamental research processes at the same time as promoting bench-to-bedside information transfer. (Anderson et al. 2009, 1)

Just a few months earlier however, a doctoral student, Nachiket Vartak, at the Max Planck Institute for Molecular Physiology in Germany published a blog post titled

Why Scientists won't use Twitter [*sic*]. Vartak seeks to explain what he sees at that time as low numbers of "natural scientists" on Twitter, stating, "scientists are substantially asocial," "rarely take laymen seriously," and prefer to communicate via e-mail. He also adds that, "Although some journals (Science, for example) offer podcasts and other 'Web 2.0' methods of disbursing information, the core process of publishing science remains tied to the Print/Paper [*sic*] method."

It is not inaccurate to claim a lack of Twitter use amongst scientists; Darling et al. (2013) estimated that only one out of every 40 scientists tweet (2.5 percent). There are two contrasting threads running through published writing and discussion about Twitter's role in science from 2009 all the way up to the writing of this chapter in early 2014: on the one hand, Twitter is revolutionizing the way scientists communicate with their peers and with the public, and the way they practice science. On the other hand, there are relatively few (or not enough) scientists using Twitter due to substantial cultural, social, and institutional barriers, and there are many reasons why scientists should use Twitter more and many ways they can and should do so. Everyone, however, seems to agree that Twitter presents possibilities for science communication in a volume and with a speed that is unprecedented, and I will examine the implications of this, beginning with peer-to-peer science communication and then moving on to a discussion of how Twitter impacts scientific public and civic engagement. Throughout I will also address questions raised and discussed by many in the literature about what, ideally, social media adoption by scientists should look like, why we aren't there yet, and how we can get there.

Twitter for Doing Science: Peer-to-Peer Communication

In the broadest sense, Twitter is useful to scientists for "mindcasting," a term evidently coined by an American professor of journalism to mean sharing information and ideas (Ben-Ari 2009). A 2009 survey of 61 scientists revealed that 91 percent of respondents prioritized using Twitter to communicate with scientific peers, and 86 percent of respondents used it to share information about their fields (Letierce at al. 2010). This can take many different forms. Scientists use Twitter to gain ready access to scientific developments and current events in science, something that can be time-consuming and overwhelming with the consistently large amount of new information and massive number of new discoveries (Ben-Ari 2009, Sack 2013; Spannangel 2011). One hydrologist tweeted in 2011 that "Twitter is useful for keeping up with current events/research more broadly than possible by journals (esp. useful for teaching)" (Vinas 2011). Scientists also share up-to-the-minute news about relevant events via Twitter, such as earthquakes, floods, and volcanic eruptions or breakthrough discoveries such as a new moon of Pluto (Ehrenberg 2012; Vinas 2011).

Twitter can be a very valuable tool for stimulating research and collaboration and increasing the size of one's scientific network. The 2009 survey cited above found that 52 percent of the scientists in the subject group use Twitter to broaden their network of other scientists. Scientists cite Twitter's usefulness for contacting other scientists in their field whom they may have never met personally, keeping in touch with colleagues from afar, and to exchange ideas for research topics and advice about methodology and equipment. The ease of networking via Twitter also seems to encourage communication and collaboration between scientists from different disciplines (Bonetta 2009; Jackson 2012; Letierce et al. 2010; Sack 2013; Spannangel 2011; Vinas 2011). The editor of a geology journal summed it up in a tweet: "Scientists can use Twitter to create/develop new collegial relationships, foster interdisciplinary research & generate ideas [*sic*]" (Vinas 2011).

In a much-cited study of Twitter and science practice, Darling et al. (2013) report that the subjects in their study (116 ocean scientist Twitter users) had an average of seven times more followers on Twitter than they had colleagues in their academic departments, and claim that 55 percent of those followers were fellow scientists and thus potential collaborators. The authors also discuss a serious concern many scientists and scholars have about exchanging new research ideas in a public social media platform: what if someone steals your ideas and publishes them before you do? Darling et al. assure their readers that tweets are dated and searchable, and thus theft of ideas or intellectual property should not be a major concern: "These social media 'time stamps' are a way to mark and share your work without the often prolonged wait times of the traditional peer-review process. Social media are at the frontier of sharing new ideas and there will undoubtedly be different opinions among users about how these tools should be used" (36).

There are indeed differing opinions on this point, and the idea of academics sharing data and research ideas via social media in the early, unpublished, or nascent stages of a study or collaboration has been fiercely debated within the scientific and academic community. It is now relatively common for attendees of professional scientific meetings to use either official (approved by or originating with the meeting's organizing body) or unofficial hashtags so that other Twitter users can follow conference proceedings. This practice has been so controversial that in 2012 the hashtag #twittergate was born out of an extensive conversation amongst academics of all kinds (not just scientists) about the ethics of live tweeting about unpublished work being presented at a professional conference (Koh 2012). Numerous bloggers weighed as well, revealing a great divide in sentiment and opinion. A philosopher blogged: " ... people who attend a conference should have the courtesy not to try to tweet the talks. If they do not have that courtesy, they should be thrown out" (Leiter 2012). Other scholars take a middle-ground, suggesting that it will and should become common practice for presenters at conferences to preface their talks by stating if they prefer the audience not to post anything about the work on social media, and that colleagues should respect those wishes (Bonetta 2009; Kolowich 2012). Still others are on the extreme opposite end of the spectrum, claiming that banning live tweeting is equivalent to banning conversations in a bar or over coffee (Warwick 2013), and that scientists are more savvy about communication than they are given credit for, being able to distinguish between the setting for discussing work, and when it isn't safe to do so (Anderson 2009).

This brings us to the larger role that social media and Twitter play in the "Open Science" and "Open Access" movements. Dan Gezelter, a chemist at the University of Notre Dame in the US and the director of the OpenScience project [*sic*] defines Open Science in a 2009 blog post as:

- Transparency in experimental methodology, observation, and collection of data.
- Public availability and reusability of scientific data.
- Public accessibility and transparency of scientific communication.
- Using web-based tools to facilitate scientific collaboration.

The more narrow Open Access movement focuses on the second of those goals, most commonly calling for the peer-reviewed science journals to allow unpaid access to scientific publications. Twitter and other social media play a powerful role here. The hashtag #openscience was first mentioned on Twitter in January 2008, and #openaccess first mentioned in November of the same year, both gradually rising in popularity with peaks in mentions at 3,527 in December 2013 and 19,849 in December 2013, respectively. The following are two representative posts using these two hashtags from the period between late 2012 to the end of 2013 (both include shortened links to longer online articles): "New study: #Openscience reaching tipping point as c.50% of papers available for free http://t.co/qAH0iaZgTl #openaccess" posted on 23 August 2013 by an EU digital agenda committee chair, and "75% of research data is never made openly available http://t.co/ zYmyMBu9 ... #opendata #openaccess #openscience" posted in December 2013 by a user who actually has the Twitter account named "@openscience." There is even a hashtag, the sole purpose of which is to allow Twitter users to exchange scientific journal articles they don't have a paid subscription for, by posting a tweet containing a Web link to the paper, the user's email address, and the hashtag #icanhazpdf, a strategy one blogger said resulted in having the PDF file of an article e-mailed to him within 20 minutes after tweeting (Hamblin 2013; Jackson 2012; Mounce 2012). This practice has been debated on science blogs, with some labeling it civil disobedience against publishers who use profit motive to bar access to crucial scientific information, and some feeling it is an immoral violation of copyright (Kroll 2011).

The conversation about social media and Open Science has revealed many deeply held values within the scientific community, also garnering sharp critiques of these values and calls to change scientific culture. One side of this argument is well summed up in a blog post by science writer David Crotty (2009), writing about why more scientists don't use social media:

> ... fear of prematurely exposing one's data and a hesitancy to offend anyone who might later be deciding on your grants, reviewing your papers or hiring

you or one of your students. Much data is hard-earned and while a vocal minority promote Open Science, the reality of it is that most researchers are very protective of the fruits of their labors. They want to feel they've fully exploited their data before releasing it to their competitors ... Scientists tend to have fairly small trusted circles ... Your preliminary data is only exposed to your labmates, perhaps to your department or a group of collaborators. It's unlikely you'll see truly open communication beyond these sorts of groups ... due to fear of committing career suicide ... Any [social] network that hopes to succeed must adapt to the culture of the community, rather than trying to rewrite it.

To support these assertions, Andeson et al. (2009) report that two of the top three concerns about adopting social media expressed by scientists at a 2008 conference were "intellectual property control [and] identity and image control." Another scientist in a 2009 blog post brings up fears about legal patent issues as well as possible widespread concerns amongst scientists about social networks lacking rigorous security (Bradley). The extreme opposite opinion is articulated succinctly in a blog post on Nature.com by Liz Neeley (2013a), the Assistant Director of Science Outreach for COMPASS, a science communication advocacy organization: "As the open science movement is demonstrating, the solitary genius of individuals is rarely superior to the speed and power of expert networks. We are stronger, wiser, and more creative as a community."

These controversies are well illustrated by a blog post written by Lior Pachter, a computational geneticist at UC Berkeley, in October 2013. Pachter writes about The GenotypeTissue Expression (GTEx) project, a genetics study funded by the US National Institutes of Health to investigate gene expression in human tissues. In the post, Pachter writes, "In a recent twitter conversation, I discovered that the [computer] program that is being used by several key GTEx consortium members to quantify the data is Flux Capacitor," and he goes into a lengthy, technical explanation of grave concerns he has about the Flux Capacitor program. The post is dated October 21, 2013, and there is a comment to the post which is time-stamped 8:59 a.m. on that same day:

We would like to thank Lior Pachter for giving us advance warning a few days ago of this post. we stand by our methods and choices and are preparing a detailed response ... Given the sensitivity of the issue and the need to consult all our collaborators ... we anticipate this response in a week to 10 days from now. Manolis Dermitzakis, University of Geneva, on behalf of GTEx and GEUVADIS consortia.

On October 31, 2013, Dermitzakis along with three other memebrs of the GTEx consoriutm, posted a lengthy response on Pachter's blog, which concluded with the statements, "We are open to constructive feedback regarding the tools we use and the analyses we perform. Finally, all data are available to any investigator who desires to perform novel analyses with their own methods and we anticipate that much improved and innovative analyses of the data will emerge with time." There are many issues embedded in this example: a government-funded, "high profile" scientific study; transparency about scientific methodology; reputations amongst colleagues; research that has a human health application. But perhaps this example boosts the point being made by Neeley (2013a) above. What if the Flux Capacitor program really isn't a good choice for this important research? Is it possible that Pachtor, via information he gained from a tweet, which he in turn posted publicly on his blog, will impact this crucial work for the better, and inspire other researchers to improve on the GTEx work by "performing novel analyses with their own methods?" This is a situation that would have never occurred without the existence of Web 2.0 social media. Is this an example of the "prompt weeding out of weak science" by social media, and the "great value in 'trials by Twitter'" (Darling et al. 2013, 39)?

Other barriers to regular social media use that have been discussed in the literature have to do with time, benefit, and reward. Scientists have expressed a simple lack of time (or unwillingness) to learning and using a new communication medium, especially when the benefits of doing so are not immediately clear. There is a feeling that time wasted on such activities would mean loss of productivity. For some, the technology itself is a barrier (Bradley 2009; Zelnio 2011). However, Darling et al. (2013) counter some of these arguments with a very real benefit to using Twitter and social media that trades in a currency all scientists understand and value: research impact. Tweeting about a scientific publication can increase exposure to one's work, and they state that some funders, as well as academic institutions for tenure and promotion decisions, are beginning to use social media as one metric for the impact of a scientist's work. The aforementioned survey of 61 scientists revealed that 86 percent of them used Twitter to widely disseminate information about their own research (Letierce et al. 2010). An evolutionary biologist and professor at UC Davis in a 2009 article in the science journal *Cell* agreed: "Twitter and regular blogging are more effective than anything else I do to publicize a paper, which was really surprising to me, If you do it right, Twitter is an effective way of telling people about your work" (Bonetta, 452).

Twitter for Outreach: Scientific Public and Civic Engagement

In a seminal 1995 paper, Baruch Fischhoff (writing about risk communication in particular) discusses the importance of communicating scientific and technical information to the lay public in ways that are respectful and take the lay audience's circumstances into consideration. He states, "People want to be treated respectfully, in addition to being leveled with. That desire is, in part, a matter of taste and, in part, a matter of power. People fear that those who disrespect them are also disenfranchising them" (142). Although some believe that "cyber-utopianism" (Obar, Zube, and Lampe 2012, 2) is a delusion (Morozov 2011), many scholars feel strongly that Web 2.0, social media, and the concomitant ready access to information and communication are now serving to level societal power imbalances (Burnett and Merchant 2011), using terms such as "mass collaboration," "transparency, peer collaboration, audience participation (Perez-Latre, Portilla, and Blanco 2011, 65; Tapscott and Williams 2006)," and "participatory cultures" (Jenkins et al. 2006, 3)." Social media and other new media technologies can foster many forms of civic engagement and create new ones (Bonzo and Parchoma 2010; Gordon, Baldwin-Philippi, and Balestra 2013). Twitter's importance as a tool in political and social movements and during times of crisis and disaster is well documented, with the use of Twitter during the Arab Spring political movements beginning in 2010 being perhaps the most well-known (Murthy 2011; Rogers et al. 2009; Yigit and Tarman 2012). In the United States, university and college students—especially African American students—used Twitter very effectively, for example, to organize political demonstrations after the highly publicized killing of African American youth Trayvon Martin (Arvin 2014; Walter Kimbrough, personal communication).

Both scientists and non-scientists have been calling for intensified and more effective science outreach and scientific civic engagement for some time, increasingly so over the last 15–20 years as the impacts of climate change and other anthropogenic perturbations are felt in natural and human systems (Lesen 2012; Poliakoff and Webb 2007; Smith et al. 2013). Professional organizations such as the American Association for the Advancement of Science (AAAS), and US funding agencies such as the National Science Foundation (NSF) are prioritizing science outreach, the NSF making it a requirement to meet a "Broader Impacts" criterion in every grant proposal (Ecklund, James, and Lincoln 2012). Poliakoff and Webb (2007) also report similar initiatives amongst U.K. funding agencies and scientific professional organizations, including organizations devoting large amounts of money to increase public appreciation about biomedical research. Jane Lubchenco, ocean scientist and former head of the US National Oceanic and Atmospheric Administration, famously called for a "new social contract for science" in a 1998 article, urging scientists to "harness the full power of the scientific enterprise in discovering new knowledge, in communicating existing and new understanding to the public and to policy-makers, and in helping society move toward a more sustainable biosphere" (495), and stating, "Scientists should be leading the dialogue on scientific priorities, new institutional arrangements, and improved mechanisms to disseminate and utilize knowledge more quickly" (496).

What Lubchenco calls for in her new social contract encompasses both scientific public engagement and scientific civic engagement as I defined these terms earlier, and necessitates science communication between scientists and the public, with policymakers being singled out as an especially important target audience. "Helping society move toward a more sustainable biosphere" and "leading the dialogue on scientific priorities, new institutional arrangements" are all embedded in my working definition of civic engagement and, as Lubchenco frames them, require scientists to work in collaboration with policymakers to develop policy, and to work within our public and private institutions to create spaces where the new scientific social contract can be supported. Social media are one of the foremost "mechanisms to disseminate and utilize knowledge more quickly;" Lubchenco's statement could thus be interpreted as a call for scientists to take full advantage of new media technology to fulfill their new social contract. I will now discuss ways the new scientific social contract can be—and is—manifested through scientists' use of Twitter, including strategies, benefits, difficulties, and challenges.

Scientific Public Outreach: "Yeah, I tried that; they kicked me out of the academy"

It is important not to take for granted that public outreach has been seamlessly incorporated into academic science practice, and the status of scientific public engagement within scientific culture and institutions bears some scrutiny. Academic scientists receive a variety of conflicting, disparate messages about science communication and outreach from peers, superiors within their institutions, the public, social scientists, and advocates for science communication and outreach. This conundrum is well summarized by Neeley (2013):

> I will never forget watching Jon Foley emphatically interrupt a discussion among young faculty with, "Your job is NOT to get tenure! Your job is to change the world." I tweeted that quip not so long ago, to which @Dreadnought1906 replied, "Yeah, I tried that; they kicked me out of the academy. Junior academics don't have the power to change the system."

Many scientists report that the attitudes of their peers, as well as academic institutions and their systems of evaluation, tenure, and promotion, do not reflect a larger cultural shift in favor of science outreach (Harley 2013). Scientific and academic communities and institutions traditionally do not reward scientists for doing work that strays outside the bounds of traditional science research, involves collaborating with social scientists, or integrates community-based methodologies and civic engagement, and doing so can risk career advancement and respect from one's scientific colleagues (Andrews at al. 2004; Bernhard 1974; Bloomfield 2005; Ellison and Eatman 2008; Mathews et al. 2005). Some researchers (Ecklund, James, and Lincoln 2012; Poliakoff and Webb 2007) have pointed out that much of the literature has been devoted to writing about how scientists could better publicly engage, rather than investigating patterns of engagement or establishing a theoretical framework for what motivates scientists to do so. I have chosen two representative studies, one from the US and one from the UK, to illustrate the current state of academic science public engagement.

Ecklund, James, and Lincoln's 2012 publication reports the results of surveys and interviews with academic biologists and physicists about their science outreach practices, attitudes, and the barriers they perceive to doing science outreach. They report on earlier work revealing that 50 percent of academic scientists "are engaged in some type of outreach … though 5 percent of the most active public scientists do half of all outreach" (1). Fifty-eight percent of their survey group engaged in outreach, and (statistically significantly) more women did so than men. More graduate students engaged in outreach than post-docs, but there was no significant difference in outreach between tenure-track and tenured faculty. When the data on scientists with and without children were analyzed separately by gender, more women with children engaged in outreach than women without children, and the same was true for men, and it is interesting to note that only three subjects mentioned that their outreach activities took place in their own children's schools. The majority of outreach activities in general, however, were with school-aged children, followed by engaging in outreach with the general public such as giving lectures or writing books aimed at lay audiences.

Poliakoff and Webb, working in the UK, focused their 2007 study on predictors of scientists' intentions to publicly engage, using the theory of planned behavior as a framework. They cite previous UK work revealing that 50 percent of the scientists studied had carried out science communication within the past year and that more than half desired to do more public engagement. Poliakoff and Webb studied a group of academic scientists from the University of Manchester, England, and found:

> … four factors influenced scientists' intentions to participate in public engagement activities over the following 12 months: … [the] extent of previous participation in public engagement activities … whether scientists regard participating in public engagement activities as positive … whether scientists feel capable of participating in public engagement activities … [and] how much scientists perceive that their colleagues are participating in public engagement activities … (254)

Taken in sum, much of the literature reports that scientists perceive and experience numerous barriers to engaging in the public realm or doing science communication, including:

- Lack of time, and internal or external pressure to spend more time on research.
- Concerns that scientists who engage are regarded poorly by their peers for doing so.
- A conviction that scientists are not adept at communication or are perceived by non-scientists to be poor communicators.
- A related belief that science communication is better done by communication experts, not scientists.
- Not seeing outreach as a legitimate part of their job or role as a scientist
- The academy, universities, and science institutions prioritize research and publications above all else, do not value science outreach, and that

scientists are not rewarded for outreach, especially in the tenure and promotion process.
- Concerns that outreach is damaging to career advancement and prestige.
- A belief that that widespread scientific ignorance amongst the public makes it too difficult to explain scientific concepts to most non-scientists.
- Public disinterest, distrust, or fear of technology, science, and scientists.
- The worry that technical vocabulary and language is too big a barrier to communication with the public. (Andrews at al. 2004; Bernhard 1974; Bloomfield 2005; Bonzo and Parchoma 2010; Ecklund, James and Lincoln 2012; Ellison and Eastman 2008; Harley 2013; Mathews et al. 2005; Neeley 2013; Poliakoff and Webb 2007; Smith et al. 2013; Zelnio 2011).

What is an academic scientist interested in public engagement and outreach to do when faced with this mountain to overcome? Ecklund, James and Lincoln (5) state that "According to our respondents, widespread change in attitude towards science outreach is difficult—if not impossible—to achieve." But they also offer a glimmer of hope, saying that the scientists they surveyed "perceive significant barriers to outreach at an individual level, within their institutions, and from the general public. And yet, though they think their departments and universities value research productivity over all else, these academic scientists still engage in outreach activities, even though they mention significant barriers to such engagement" (5). If barriers to science outreach are breaking down, it seems to be a slow process. However, social media offers new avenues for doing public engagement, and I will now, finally, turn my attention to Twitter and its role in scientific public outreach.

What Can Scientists Communicate in 140 Characters?

For the remainder of this section of the chapter, I will use the terms "science communication" as shorthand for communication between scientists and the public rather than peer-to-peer communication, which I have already discussed. Science communication of this type is most often framed as scientists translating scientific information and knowledge to non-scientists, and as Fischhoff (1996) points out, ideally in a way that is respectful, understandable, equitable, and accurate. Because of Twitter's ease of access for anyone who has the correct technology, some feel it offers unprecedented opportunities for two–way exchanges between scientists and the public: "… scientists and non-scientists communicate and collaborate in transdisciplinary teams. Both groups profit from each other: Non-scientists have direct access to the latest scientific results …" (Spannangel 2011, 1). One scientist makes the strong and simple claim on his blog that "If you, as a scientist, don't use [Twitter], you are putting yourself at a big disadvantage" (Knoepfler 2012).

Many scientists have grave concerns that 140 character tweets are simply too short to explain a scientific concept, that there is no room for context, and this leaves a scientist vulnerable to her followers misunderstanding and misinterpreting meaning

(Ehrenberg 2012). Sandu in a 2011 blog post states with conviction that, although scientists gain other benefits from its use, "social media is not the place to 'explain' science." Bonetta (2009) quotes one scientist say saying, "It is a double-edged sword. The majority of my tweets are pointers to other resources, so there is a headline ... and a link to the resource. You don't need more than 140 characters for that. However you cannot have a decent, full-blown, high-level scientific debate via Twitter." This is why, although some in the literature claim Twitter is the most popular Web 2.0 social medium, some claim it is blogs: "Blogs remain the social medium of choice for science, as they have space to expound complex arguments but retain an immediacy that invites fruitful discussion" (Vence 2013).

Others see the 140 character limit as an advantage. A detailed Twitter-for-science instruction manual states, "Reaching out to external audiences is something that Twitter is exceptionally good for. Making links with practitioners in business, government, and public policy can happen easily. Twitter's brevity, accessibility and immediacy are all very appealing to non-academics" (Mollett, Moran, and Dunleavy 2011, 7). At least one climate scientist feels Twitter is an opportunity for scientists to prevent miscommunication and misinterpretation because it is a rare instance when scientists can connect directly with large numbers of the public, tweeting that "climate scientists in particular should use Twitter because too much misquoting by 3rd parties with political agendas on both sides [*sic*]" (Vinas 2011).

Scientists have created communities in the "Twitterverse" through the use of hashtags. For those who are interested in the Open Science and Open access movements, there are the aforementioned #openaccess and #openscience. The hashtag #scicomm, for science communication, is particularly active, especially amongst Twitter users from the UK. The probable first tweet using the hashtag #scicomm was posted on January 16, 2009 by Jo Brodie @jobrodie, the self-described "Public Engagement Co-ordinator" at the University of College London's Computer-Human Interaction for Medical Devices (CHI+MED) program, and there have been months since then where there were over 2,400 original posts (not replies or re-tweets) using #scicomm (the hashtag #sciencecommunication, likely taking up too many valuable characters in a tweet, has not been very popular, the same being true for #civicengagement). There are also somewhat popular hashtags specific for the "ScienceOnline" community. On its website, ScienceOnline (http://scienceonline.com) states that its "mission is to cultivate the ways science is conducted, shared, and communicated online." ScienceOnline encompasses a diverse community of people whose work or interests in some way overlap with the biophysical sciences, and a major thrust is science communication of all kinds. ScienceOnline hosts online forums, projects, and conferences (ScienceOnline Together, ScienceOnline Climate, ScienceOnline Teens), all of which generate traffic on Twitter using hashtags. #ScioX (for ScienceOnline in general, and #ScioX14 for its 2014 conference, for example); #ScioAdvocacy for the subset of people in the ScioX community who are interested in science advocacy; #ScioOceans and #ScioClimate for the ScioX community interested in ocean science, climate science, or the yearly ScienceOnline meetings devoted to those topics). In the use of these hashtags, people can exchange information and use Twitter for science practice, but can also use Twitter as a place where they can find each other, support each other, and exchange information about science outreach itself.

Humanizing Scientists and Making Science Transparent

There is general agreement in the literature and in online writing that one of Twitter's greatest social impacts is that it gives the lay public a chance to learn about the daily life and work of scientists directly from the scientists themselves. This is viewed by most as an important paradigm shift in science communication, one that could alleviate the dehumanized way science is often portrayed in media and popular culture, and this could help scientists gain greater trust from the public. Ecklund, James and Lincoln (2012) found that many of their scientist respondents felt that the American public distrusts scientists. Gordon, Baldwin-Philippi, and Balestra (2013), in their literature review on digital media, human behavior, and civic engagement, cite trust as a main factor determining civic engagement and call for more research into the impact social media and digital networks have on how much people trust each other. In a COMPASS blog, Neeley (2013b) discusses the work of Princeton University researcher Susan Fiske, who found during her research that scientists are respected but not trusted by the public, are seen as "competent but cold," and that "[t]rustworthiness … is a quality produced by a combination of perceived warmth and competence. Warmth in this work is not exactly 'likeable,' rather, it refers to the judgments we make about person's motives. Competence is their ability to act on those intentions." Those in the Open Science movement call for more transparency from scientists about their methods (Gezelter 2009).

Via a sampling of blog post quotes below, one gleans that some scientists are quite hopeful about the power of Twitter in this arena, especially in science education:

- One thing everyone agrees with is that scientists have to learn to communicate their work to non-scientists. Twitter allows anyone to see science in a way that is more accessible, such as scientists reporting on their daily failures and successes (Bonetta 2009).
- Twitter is also a great way to reach out to the general public and give back. [You can] share tales from your research (or more personal stories that demonstrate that scientists are human too) … (Jackson 2012).
- … we've found yet another reason for you to start using Twitter in the classroom: It offers users a window into the lives, work and perfunctory musings of some of the most important contemporary intellectuals in the world (Cameron 2013).
- [writing about high school students interacting with a scientist via Twitter and Skype]: Simon agreed to Skype with the class and we had one of the best hours of my teaching career. The students asked Simon questions,

interacted with him, and saw him a as a regular person—someone like them (Russo and Romano 2013).

Two of the best examples of Twitter being used in this way are the hashtags #overlyhonestmethods, which "went viral" in January 2013 (Lorch 2013; Sandle 2013), and #sciconfessions during summer 2013 (Donald 2013). The hashtag #overlyhonestmethods began with a tweet from a self-described neuropharmacologist who posted several tweets using that hashtag on January 7, 2013 including, "incubation lasted three days because this is how long the undergrad forgot the experiment in the fridge #overlyhonestmethods." The hashtag is still in use as of this writing in early 2014. Science bloggers Mark Lorch and Time Sandle (2013) wrote about the phenomenon and reprinted several choice tweets using the hashtag:

> "… the chemicals were combined & stirred by hand for 2 hours by our project students as they were getting on our nerves" … "The experiment was left for the precise time that it took for us to get a cup of tea" … "the eppendorf tubes were 'shaken like a polaroid picture' until that part of the song ended" … "Blood samples were spun at 1500rpm because the centrifuge made a scary noise at higher speeds #overlyhonestmethods"

Lorch also wrote:

> So what started as a single tweet from a frustrated scientist has ended up becoming one of the most fabulous, frank and funny pieces of science communication I've seen in a long time. Some might worry that these tweets have presented scientists as hapless and undermined confidence in science. But I think they have provided a rare insight into the everyday lives of scientists and demonstrated that we are human like everyone else. Moreover, #overlyhonestmethods has managed to demystify science in a way that no other example of science reporting, blogging or broadcasting can quite manage.

The hashtag #sciconfessions was first used in summer 2010 and seems to have been particularly active in late summer 2013 (Donald 2013; Topsy 2014). Tweets using the #sciconfessions tag give reports of research methods that have gone wrong. Donald (2013) both summarizes and quotes several of the tweets:

> … mouth pipetting bacterial cultures … travelling home by bus after spilling some vile-smelling compound on one's trousers … "I have fallen asleep operating a $250 million dollar telescope …" "I have eaten an organism I couldn't ID just to get rid of it …" what strikes me most about the images conjured up by these 140 character vignettes is the fact that, despite the too frequent portrayals of scientists as somehow inhuman in film and TV, we are on the contrary a very human bunch. We mess up, in big and small ways, we laugh at ourselves, we talk

to ourselves, our bacteria and even our apparatus. we may dance in the lab when no one's looking and we worry about our image.

@realscientists is a Twitter account that rotates among different scientists, science policy makers, and science communicators each week. According to the RealScientists (2013) blog, following @realscientists on Twitter helps one "Get a feel for what scientists are actually doing locked away in those labs, and checkout all the things you can do with a science degree. It's Science by Twitter—not so sinister or secret now …"

Public Participation and Empowerment Through Social Media

Digital media are transforming scientific public engagement through enabling members of the lay public new ways to participate in scientific discourse, scientific debates, and in the practice of science itself. Paulos, Kim and Kuznetsov (2011) describe projects they did using SMS, mobile sensing, and mobile-phone technology to give citizens opportunities to collect and gain access to air quality data in real time. They gave air quality sensors to taxi drivers and students in Accra, Ghana, and reported that as a result of being able to access this information at any time, the participants developed strategies for traveling around the city to minimize their own exposure to poor air quality. They also worked with the City of San Francisco to attach air quality sensors to street sweepers. Taken as a whole, they feel their work demonstrates the power of technology to develop a strategy where "'participation' in data collection is interpreted broadly as citizen, civic, and infrastructure" (176).

There are many other examples from around the world of scientists using social media and digital and mobile technology to increase citizen participation in science. Anderson et al. (2009, 1–2) explain that, "Examples of user-generated and manipulated content in science range from open notebook science … and social networks for researchers or lab groups … to crowdsourcing (also known as volunteer or citizen science) … clinical trial subject recruitment … and research findings dissemination and adoption … " A Canadian research team is asking citizens to report the snow depth in their area (from anywhere in the world) using the hashtag #snowtweets (http://snowcore.uwaterloo.ca/snowtweets/). Three organizations have collaborated to develop a program using several hashtags for hikers to tweet pictures of wildfire damage and recovery in Mt. Diablo State Park in California (http://nerdsfornature.org/monitor-change/diablo.html). Twitter is also a powerful collaborative tool for advocacy, just one example being a 2011 blog reporting that "Abby Kavner, a mineral physicist, agrees that Twitter connects and mobilizes people around science topics … 'last year's saving of California's state rock serpentinite was definitely a Twitterdependent victory'" (Vinas 2011). Preece and Schneiderman (2009, 25) caution us that metrics need to be developed in order to evaluate the real-world impacts these types of projects may or may not have: "Can highly active healthcare discussion groups lead to lower mortality or morbidity? Can community crime reporting applications create safer neighborhoods? Can international development or world peace be advanced by social media? Collecting activity metrics and studying efficacy will require advances in data analysis, statistics, and visualization tools ..."

Loss of Control and Prestige?

While many celebrate social media's role in humanizing scientists, increasing the transparency of science practice, and advancing in public empowerment and citizen science, there are some who are not as enthusiastic. Some scientists have expressed worries that their peers see engagement in social media as exhibiting a lack of focus and a departure from traditional academic practice, and many pre-tenure scientists are counseled by their tenured peers to behave conservatively (Harley 2013; Vence 2013). Some believe that a fear of loss of control is at the core of these conflicts. Sociologists and philosophers of science have posited that many of the norms within the scientific community exist to maintain a system of evaluating their peers in order to protect, control, and legitimize the position of science and scientists in society (Mulkay 1991; Owen-Smith 2001). Use of social media by scientists can be seen as challenging those norms. Warwick (2013) writes that "[s]ocial media involves a loss of control and an exercise in trust and openness." Yeo et al. (2012) write that "[on Twitter], lay audiences can receive, repost, and comment on new scientific findings. As the definitions of expert and public communication continue to change, and the media environment and public audiences adapt to it, scientists will have no choice but to evolve, too." Fullick (2012) feels strongly that power, control, and the desire to protect social prestige is at the core of the controversy amongst scholars about live-tweeting conferences and the like:

> One of the key problems brought up in the online debate has been that of determining what knowledge is public and what's private, and who gets to decide how dissemination of that knowledge happens (where, and when, and who the audience will be), who has the "right" to share ideas. In my opinion, control is one of the fundamental elements of this discussion ... the connection between access and control. Control is also exercised through authority, which is closely tied to expertise and peer recognition. So we see some scholars reasserting a form of academic credibility by putting down other scholars as mere opportunists, not "real" academics. In this way the boundary between "academic" and "anything else" is redrawn by those who are already inside it–or those who hope to be allowed in.

These issues—the scientific community's concepts of expertise, control, access, and prestige—certainly need to be addressed explicitly as scientists grapple with emerging formulations of their roles and responsibilities under a new social contract. Scientists and other academics have less and less power to limit access to knowledge as information is exchanged with ease, mediated by social networks and digital media. This may challenge many scholars' very sense of who they are. These questions will have to be debated between those who applaud these changes, those who fear and abhor these changes, and those in between, as the very practice of science and academics shifts in the face of this new communication paradigm.

Conclusions

Through this literature review, I have demonstrated that Web 2.0 and social media offer a host of novel opportunities for science communication, science outreach, and scientific public and civic engagement. In fact, the very essences of civic engagement and communication in all forms are evolving as new digital media technology is invented and adopted worldwide. Twitter and blogging are having the most impact, with the highest adoption rates amongst scientists, creating new venues for peer-to-per communication. The way science is practiced is changing as a result, with new possibilities for forging collaborations, as well as creating and disseminating scientific knowledge. Twitter also allows scientists to directly engage the public with an ease and in numbers as never before, creating the potential for social media to remake these relationships and interactions, allowing a more reciprocal exchange of knowledge and information, and giving members of the public innovative means for participation in science. Social media also represent a vehicle for science to be more open and transparent to the public in methodology, availability, and communication. This in turn creates a space where the public can witness scientists' in their work and lives, humanizing scientists and possibly increasing public trust in science.

This new paradigm for scientific outreach and communication is in step with current appeals for scientists to play greater social roles and deepen their participation in the civic realm. At the same time, despite outreach initiatives from some prominent scientific professional organizations and funding agencies, academic institutions and norms within the scientific community are not congruent with a "new social contract for science." A significant proportion of scientists value and are engaged in public outreach and science communication, despite serious barriers to doing so, including lack of reward for outreach in tenure and promotion decisions, lack of support for outreach amongst peers, and lack of training in outreach and communication.

There are a vast and growing number of resources for scientists who want to publicly engage, including funding, training, organizations, and online networks, but real change is needed to empower scientists to make communication and outreach an integral part of scientific practice. Academic institutions must include public engagement—including metrics for participation and presence in social media—in the way scientists are evaluated and rewarded, especially in tenure and promotion decisions. Training in outreach and science communication should be a regular part of science higher education and scientific professional development.

However, there are gaps in our understanding of scientific public engagement that make it impossible to make truly informed recommendations about how to address some of these challenges. More research needs to be done to study not only the patterns of scientific public engagement (who is engaging, how they are engaging, and why they are engaging), and which scientific public outreach strategies are most effective (how best to engage) but, as Poliakoff and Webb (2007) suggest, the beliefs and values scientists hold that influence their behavior and decisions about engagement and outreach. Especially as digital media enable greater transparency in science and greater knowledge access for the public, scientists as a community must candidly deliberate amongst themselves about deeply embedded values, attitudes, and practices regarding their status in society, and conscious or unconscious attempts to protect their prestige. The nature of and definition of expertise requires debate and reflection by scientists as well: What is an expert, academic or a scientist? Who is allowed into that group? Who should have access to information and knowledge, and why? It is not until scientists have more honest and clear answers to those questions that a new model for science communication and public engagement can emerge. This sea change must happen before scientists can take full advantage of the outreach possibilities offered by digital social media.

References

Anderson, Kent. 2009. Scientists Are Using Social Media Tools (and May Be Using Social Networks, Too). Posted on November 3, 2009, scholarlykitchen. sspnet.org/2009/11/03/scientists-are-using-social-media-tools-and-may-be-using-social-networks-too/.

Anderson, P.F., J. Blumenthal, D. Bruell, M. Rosenzweig, M. Conte, and J. Song. 2009. An online and social media training curricula to facilitate bench-to-bedside information transfer. In *Positioning the Profession: the Tenth International Congress on Medical Librarianship*: pp. 1–11. http://www.academia.edu/2841518/An_online_and_social_media_training_curricula_to_facilitate_bench-to-bedside_information_transfer.

Andrews, E., A. Weaver, D. Hanley, J.H. Shamatha, and G. Melton. 2004. Scientists and public outreach: participation, motivations, and impediments, http://cires.colorado.edu/education/outreach/rescipe/papers/andrewsJGE2005preprint.pdf.

Arvin, C. 2014. 'Clicktivism' Moves Civil Rights Forward in a New Generation. Posted on March 14, 2014, http://www.blackvoicenews.com/news/49416-clicktivism-moves-civil-rights-forward-in-a-new-generation.html.

Ellison J., and T.K. Eatman. 2008. Scholarship in public: knowledge creation and tenure policy in the engaged University. *Imagining America*, Syracuse, http://

imaginingamerica.org/fg-item/scholarship-in-public-knowledge-creation-and-tenure-policy-in-the-engaged-university/.

Ben-Ari, E. 2009. Twitter: What's All the Chirping About?. *BioScience*, 59 (7): 632.

Bernard, H.R. 1974. Scientists and Policy Makers: An Ethnography of Communication. *Journal of Human Organization*, 33(3): 261–76.

Bloomfield, V. 2005. Civic Engagement and Graduate Education. *Communicator*, 38:3.

Bonetta, L. 2009. Should you be tweeting? *Cell*, 139(3): 452–3.

Bonzo, J., and G. Parchoma. 2010. The paradox of social media and higher education institutions. In *Proceedings of the 7th international conference on networked learning*, eds L. Dirckinck-Holmfeld, V. Hodgson, C. Jones, M. de Laat, D. McConnell, and T. Ryberg, 912–18. http://www.lancaster.ac.uk/fss/organisations/netlc/past/nlc2010/abstracts/PDFs/Bonzo.pdf.

Bradley, David. 2009. Gen-F Scientists Ignoring Social Networking. Posted on October 7, 2009, www.sciencebase.com/science-blog/gen-f-scientists-ignoring-social-networking.html.

Burnett, C., and G. Merchant. 2011. Is there a space for critical literacy in the context of social media?. *English Teaching: Practice & Critique*, 10(1): 41–57.

Burns, T.W., D.J. O'connor, and S.M. Stocklmayer. 2003. Science Communication: A Contemporary Definition. *Public Understanding of Science*, 12: 183–202.

Cameron, Karen. 2013. 100 Scientists. 140 characters away: Using Twitter in the classroom. Posted by on April 3, 2013, Classroom 2.0, www.classroom20.com/forum/topics/100-scientists-140-characters-away-using-twitter-in-the-classroom.

Cook, Gareth, 2011. Science and Twitter #mixwell. *Boston Globe* Op.-Ed, Posted October 2, 2011, http://www.boston.com/bostonglobe/editorial_opinion/oped/articles/2011/10/02/science_and_twitter_mixwell/.

Crotty, David. 2009. Scientists Still Not Joining Social Networks. Posted on October 19, 2009, http://scholarlykitchen.sspnet.org/2009/10/19/scientists-still-not-joining-social-networks/.

Darling, Emily S., David Shiffman, Isabelle M. Côté, and Joshua A. Drew. 2013. The role of Twitter in the life cycle of a scientific publication. *Ideas in Ecology and Evolution*. 6: 32–43.

Donald, Athene. 2013. Scientists confess via Twitter: A Twitter hashtag has provoked an illuminating string of admissions from scientists, Posted on August 5, 2013, Occam's Corner, Hosted by the Guardian, www.theguardian.com/science/occams-corner/2013/aug/05/people-in-science.

Ecklund E.H., S.A. James, and A.E. Lincoln. 2012. How Academic Biologists and Physicists View Science Outreach. *PLoS ONE*, 7(5): e36240.

Ehrenberg, Rachel. 2012. Scientists embrace Twitter for spreading the word and hashing through new data. Posted on October 15, 2012, Science News Online, https://www.sciencenews.org/article/scientists-embrace-twitter-spreading-word-and-hashing-through-new-data.

Ehrlich, T., ed. 2000. *Civic responsibility and higher education.* American Council on Education Oryx Press Series on Higher Ed: Rowman & Littlefield Publishers.

Federer, L. 2013. Uses for Twitter across disciplines and throughout the scientific process. *Ideas in Ecology and Evolution,* 6(1): 44–5.

Fischhoff, B. 1995. Risk perception and communication unplugged: Twenty years of process. *Risk analysis,* 15(2): 137–45.

Fullick, Melonie. 2012. Tweeting out loud: ethics, knowledge and social media in academe. Posted on October 16, 2012, Impact of Social Sciences blog, blogs. lse.ac.uk/impactofsocialsciences/2012/10/16/fullick-tweeting-out-loud/.

Gezelter, Dan. 2009. What, exactly, is Open Science? Posted on July 28, 2009, http://www.openscience.org/blog/?p=269.

Gordon, E., J. Baldwin-Philippi, and M. Balestra. 2013. Why We Engage: How Theories of Human Behavior Contribute to Our Understanding of Civic Engagement in a Digital Era. *Berkman Center Research Publication,* (21). https://cyber.law.harvard.edu/publications/2013/why_we_engage.

Hamblin, Stephen. 2013. Social Media and Academics … Posted on June 25, 2013, winawer.org/blog/2013/06/25/social-media-and-academics/.

Harley, D. 2013. Scholarly Communication: Cultural Contexts, Evolving Models. *Science,* 342: 80–82.

Jackson, Morgan. 2012. Twitter for Scientists (and why you should try it) (#ScienceShare). Posted on Jan 2, 2012, www.biodiversityinfocus.com/blog/2012/01/02/twitter-for-scientists-and-why-you-should-try-it-scienceshare/.

Jenkins, Henry. 2006. *Convergence culture: Where old and new media collide,* New York: NYU Press.

Jenkins, Henry, Ravi Purushotma, Margaret Weigel, Katie Clinton, and Alice J. Robison. 2006. Confronting the Challenges of Participatory Culture: Media Education for the 21st Century. An Occasional Paper on Digital Media and Learning. *John D. and Catherine T. MacArthur Foundation.* http://mitpress.mit.edu/sites/default/files/titles/free_download/9780262513623_Confronting_the_Challenges.pdf.

Koh, Adeline. 2012. #Twittergate: What are the ethics of livetweeting at conferences? (Exact date of posting unkown), storify.com/adelinekoh/what-are-the-ethics-of-live-tweeting-at-conference.html.

Kolowich, Steve. 2012. Scholars debate etiquette of live-tweeting academic conferences. Posted on October 2, 2012, www.insidehighred.com/news/2012/10/02/scholars-debate-etiquette-live-tweeting-academic-conferences.

Knoepfler, Paul. 2012. The scientist's top 10 guide to Twitter. Posted on May 31, 2012, Knoepfler Lab Stem Cell Blog, www.ipscell.com/2012/05/the-scientists-top-10-guide-to-twitter/.

Kroll, David. 2011. #icanhazpdf: Civil disobedience? Posted on December 22nd, 2011, http://cenblog.org/terra-sigillata/2011/12/22/icanhazpdf-civil-disobedience/.

Leiter, Brian. 2012. Tweeting conferences? Posted on October 02, 2012, Lieter Reports: A Philosophy Blog, leiterreports.typepad.com/blog/2012/10/tweeting-conferences.html.

Lesen, Amy E. 2012. Oil, floods, and fish: the social role of environmental scientists. *Journal of Environmental Studies and Sciences*, 2(3): 263–70.

Letierce, Julie, Alexandre Passant, John Breslin, and Stefan Decker. 2010. Understanding how Twitter is used to spread scientific messages. In: *Proceedings of the WebSci10: Extending the Frontiers of Society On-Line*, April 26–27th, 2010, Raleigh, NC: US. http://journal.webscience.org/314/2/websci10_submission_79.pdf.

Lorch, Mark. Scientists take to Twitter to reveal their less than scientific methods. Posted on January 10, 2013, theguardian.com, www.theguardian.com/science/blog/2013/jan/10/scientists-twitter-methods.

Lubchenco, Jane. 1998. Entering the century of the environment: a new social contract for science. *Science*, 279(5350): 491–7.

Mathews, D.J.H., A. Kalfoglou, and K. Hudson. 2005. Geneticists' Views on Science Policy Formation and Public Outreach. *American Journal of Medical Genetics*, 137A: 161–9.

McCormick, T. H., H. Lee, N. Cesare, and A. Shojaie. 2013. Using Twitter for Demographic and Social Science Research: Tools for Data Collection. In *Sociological Methods and Research, Population Association of American Annual Meeting*, New Orleans, April 11–13, 2013.

Mollett, A., D. Moran, and P. Dunleavy. 2011. Using Twitter in university research, teaching and impact activities: A guide for academics and researchers. *London School of Economics and Political Science: LSE Public Policy Group*. http://blogs.lse.ac.uk/impactofsocialsciences/files/2011/11/Published-Twitter_Guide_Sept_2011.pdf.

Morozov, E. 2012. *The net delusion: The dark side of Internet freedom.* PublicAffairs; Reprint edition.

Mounce, Ross. Twitter tips for Systematists. Posted on January 11, 2013, rossmounce.co.uk/2013/01/11/twitter-tips-for-systematists/.

Neeley, Liz. 2013a. Science Communication at a Tipping Point. Posted on May 15, 2013, Nature.com, Soapbox Science, blogs.nature.com/soapboxscience/2013/05/15/science-communication-at-a-tipping-point.

Neeley, Liz. 2013b. Is "Cold But Competent" a Problem in Science Communication?. Posted on October 21, 2013, Compassblogs:Answering "So What?" in science communication, compassblogs.org/blog/2013/10/21/is-cold-but-competent-a-problem-in-science-communication/.

Obar, J.A., P. Zube, and C. Lampe. 2012. Advocacy 2.0: An analysis of how advocacy groups in the United States perceive and use social media as tools for facilitating civic engagement and collective action. *Journal of Information Policy*, 2: 1–25.

RealScientists. 2013. "About @RealScientists," Real science, from real scientists, science communicators, writers, artists, clinicians, Posted February 2013, realscientists.wordpress.com/about/.

Pachter, Lior. 2013. GTEx is throwing away 90% of their data. Posted on October 21, 2013, Bits of DNA: Reviews and commentary on computational biology, http://liorpachter.wordpress.com/2013/10/21/gtex/.

Response to: "GTEx is throwing away 90% of their data." Posted on October 31, 2013 by Manolis Dermitzakis, http://liorpachter.wordpress.com/2013/10/31/response-to-gtex-is-throwing-away-90-of-their-data/.

Paulos, E., S. Kim, and S. Kuznetsov. 2011. The Rise of the Expert Amateur: Citizen Science and Microvolunteerism. In *Social Butterfly to Engaged Citizen: Urban Informatics, Social Media, Ubiquitous Computing, and Mobile Technology to Support Citizen Engagement*, eds M. Foth, L. Forlano, C. Satchell, and M. Gibbs, 167–96. Cambridge, MA: MIT Press.

Pérez-Latre, F.J., I.P. Blanco, and C. Sanchez. 2011. Social Networks, Media and Audiences: A Literature Review. *Comunicación y sociedad*, 24(1): 63–74.

Poliakoff, E., and T.L. Webb. 2007. What factors predict scientists' intentions to participate in public engagement of science activities?. *Science communication*, 29(2): 242–63.

Preece, J., and B. Shneiderman. 2009. The reader-to-leader framework: Motivating technology-mediated social participation. *AIS Transactions on Human-Computer Interaction*, 1(1): 13–32.

Rogers, R., M. Stevenson, and E. Weltevred. 2009. Social research with the web. Global information society watch, http://www.govcom.org/publications/full_list/GISWatch_DMI.pdf.

Russo, Cristina, and John Romano. 2013. Guest Post: Creating scientists in 140 characters. Posted September 9, 2013, Plos Blogs, http://blogs.plos.org/scied/2013/09/09/guest-post-creating-scientists-140-characters/.

Sack, Georgeann. 2013. Calling All Scientists: Use Twitter like a pro, for your own good. Posted on June 6th, 2013, Active Scientist blog, www.activescientist.com/calling-all-scientists-use-twitter-like-a-pro-for-your-own-good.

Sandle, Tim. 2013. Scientists use Twitter as a medium to be 'open and honest.' Posted on January 12, 2013, Digital Journal, www.digitaljournal.com/article/341140.

Sandu, Oana. 2011. How much science in a tweet? Posted on March 22, 2011, http://astronomycommunication.com/2011/03/22/how-much-science-in-a-tweet/.

Spannangel, Christian. 2011. How an open scientist can use Twitter. Euroscience working group "Science Communication." https://www.euroscience.org/tl_files/Euroscience/Activities/Workgroups/Science%20Communication/Tip%20Sheets/Tip_sheet_twitter_Spannagel.pdf.

Smith, Brooke, Nancy Baron, Chad English, Heather Galindo, Erica Goldman, Karen McLeod,Meghan Miner, and Elizabeth Neeley. 2013. COMPASS: Navigating the rules of scientific engagement. *PLoS Biology*, 11(4): e1001552.

Tapscott, Don, and Anthony D. Williams. 2008. *Wikinomics: How Mass Collaboration Changes Everything*, New York: Penguin.

Tarman, B., and M.F. Yigit. 2012. The Impact of Social Media on Globalization, Democratization and Participative Citizenship. *JSSE-Journal of Social Science Education*, 12(1): 75–80.

Vartak, Nachiket. 2009. Why Scientists won't use Twitter. Posted on February 15, 2009, The Daily Nash-On Blog, nachiket.wordpress.com/2009/02/15/why-scientists-wont-use-twitter/.

Vence, Tracy. 2013. Science Gone Social: From keeping up with the literature to sparking collaborations and finding funds, scientists are storming social media. Posted on June 27, 2013, Genetic Engineering and Biotechnology News Blog, www.genengnews.com/keywordsandtools/print/3/31788/.

Viñas, Maria José. 2011. Why should scientists use Twitter? Posted on July 20, 2011, http://blogs.agu.org/sciencecommunication/2011/07/20/why-scientists-use-twitter/.

Warwick, Claire. 2013. The terror of tweeting: social medium or academic message? Posted on February 5, 2013, Higher Education Network, Guardian Professional, www.theguardian.com/higher-education-network/blog/2013/feb/05/academic-twitter-technology-social-media-universities?CMP=twt_gu.

Yeo, Sara K., Dominique Brossard, Dietram A. Scheufele, Paul Nealey, and Elizabeth A. Corley. 2012. Tweeting to the Top. Posted on July 2, 2012. The Scientist, www.the-scientist.com/?articles.view/articleNo/36274/title/Opinion--Tweeting-to-the-Top/.

Zelnio, Kevin. 2011. On Naïveté Among Scientists Who Wish to Communicate. Posted October 4, 2011, Permanent Address: http://blogs.scientificamerican.com/evo-eco-lab/2011/10/04/on-naivete-among-scientists-who-wish-to-communicate/.

Index

Milton Keynes UK
Ingram Content Group UK Ltd.
UKHW031151141024
449569UK00024B/880